WITHDRAWN

EXPERIMENTAL
BIOLOGY

PRENTICE-HALL BIOLOGICAL SCIENCE SERIES
William D. McElroy and Carl P. Swanson, Editors

EXPERIMENTAL BIOLOGY

second edition

Richard W. Van Norman

Professor of Biology
University of Utah

Prentice-Hall, Inc., Englewood Cliffs, New Jersey

Current printing (last digit): 10 9 8 7 6 5 4 3 2

13–294710–2

Library of Congress Catalog Card Number 74–105444

Printed in the United States of America

PRENTICE-HALL INTERNATIONAL, INC., *London*
PRENTICE-HALL OF AUSTRALIA, PTY. LTD., *Sydney*
PRENTICE-HALL OF CANADA, LTD., *Toronto*
PRENTICE-HALL OF INDIA PRIVATE LIMITED, *New Delhi*
PRENTICE-HALL OF JAPAN, INC., *Tokyo*

Modern biology is the product of years of evolution. Many problems have been solved by the classical descriptive approach. More and more, the remaining questions are intimately concerned with the responses of the individual cell, or with subtle interrelationships. These problems remain because they are difficult to observe. It has become increasingly important to rely upon quantitative observations and upon the tools provided by chemistry, physics, and mathematics.

The great middle ground between the biological and physical sciences can be approached from either side. Biologists continue the attack using their own methods. Numerous competent physical scientists have been challenged by the intricacy of the living organism. Unfortunately, communication between these two groups is often difficult, but they are beginning to learn each other's language. The experimental approach is the same: only the philosophies are different.

A fraction of the failure of communications between the biological and physical sciences rseults from impressions or prejudices developed during the training period. This book is an attempt to introduce undergraduate biology and physical science majors to each other at an early age, before these preconceptions have a chance to develop. The chief hope is that young biologists can learn to appreciate the physical approach to problems, and that young physical scientists can learn some of the fascination of biological research.

The book originally grew out of a course that is offered at the University of Utah. This course in Experimental Biology is itself an experimental introduction to biological research. Experience has taught us that undergraduates *can* apply the experimental approach to biological problems. This book was written to help make the learning processes faster and easier.

There is no intention of assembling a set of instructions which, if scrupulously followed, will guarantee success in research. Research must be a creative, personal process. In our investigations, we learn by our failures as well as by our successes. Therefore, the most effective training program is a number of years of experience. On the other hand, young minds are agile and imaginative and full of potential contributions. A general guide may help these minds to produce sooner, to avoid some of the failures, and to escape from the necessity of learning everything the hard way. It is imperative that laboratory work accompany the discussions presented here. The bibliography lists several good guides that will help the student learn some of the intimate details of laboratory experience. After starting under a competent teacher, the student will wish to pursue problems on his own initiative. At the end of several of the chapters, I have suggested some problems that he might undertake when he reaches that stage.

This book is intended to be a starting point rather than an end in itself. With this thought in mind, most of the chapters have references to more advanced discussions. All the works cited in the first edition have been re-examined. Some have been retained. A number of exciting new works have appeared in the intervening years. Some of the advanced discussions referred to will be very difficult for the usual bright college sophomore to read. By introducing the subject in simple terms, perhaps this book can make the more advanced discussions easier.

During the preparation of the first edition, I rearranged the table of contents, that is, the order of the chapters, several times. The problem is that the various topics are so closely interrelated that it is difficult to speak of one subject without also mentioning the others. Experimental Design (Chapter 17) might come first, for example, but the subject is difficult and can best be illustrated by examples which will be unfamiliar at the beginning. The best organization, apparently, would be to start in the middle and proceed in all directions. When the time came to prepare the present edition, I could see no very good reason to change the original order; however, I have added one new chapter.

The final schedule of topics, then, is a compromise. The first few chapters form a general introduction to quantitative, experimental biology. The next major section is devoted more specifically to those techniques and instruments which are such an important part of the everyday work of the laboratory. Finally, the data obtainable by such techniques are treated mathematically and statistically, and converted into something meaningful and useful.

Many kind people have assisted in the preparation. A few sharp-eyed former students may recognize their own data in some of the examples. Many of these same students were among the best critics of most of the chapters.

Several competent teachers—who choose to remain anonymous—made many valuable suggestions after using the first edition. My colleagues at the University of Utah read some of the chapters and discussed a great many ideas with me. They do not know how much they helped, and I hesitate to name them for fear that the faults that remain may be attributed to them—that responsibility is mine alone. Professors Lorance Greenlee, E. W. Hanly, Robert R. Kadesch, James L. Lords, Ivan M. Lytle, I. B. McNulty, W. W. Newby, John D. Spikes, and W. N. Strickland deserve my special thanks. My most generous and persistent helper was Dolores K. Van Norman, who contributed more than she knew. In addition to helping with the typing and assembling of the final manuscript, she read, criticized, encouraged, and assisted in a great many other ways.

Richard W. Van Norman
Salt Lake City, Utah

CONTENTS

**EXPERIMENTAL
BIOLOGY**

SCIENCE

Biology has become a quantitative, experimental science. So much emphasis is now being placed upon the use of mathematical, physical, and chemical tools in the attempt to learn about living things that the beginner may be bewildered. It is the purpose of this book to provide a starting point from which modern research in biology can be explored.

Science surrounds us in modern America. Every day we are confronted with some "scientific" marvel. Today's children probably know more about space flight, radar, and viruses than science itself knew a generation ago. Science is prominent in all our lives, and it is the duty of any educated person to find out what science is about.

As a beginning, therefore, one should ask such questions as: What is science? What are the goals of scientific activity? What is research? Is all of science devoted to improving human comfort and convenience? What is the difference between the scientist and an "ordinary man"? By what methods are the goals of science achieved? Does science have any limitations? Let us briefly explore these and some other questions.

WHAT IS SCIENCE? Science has been with us for a very long time, and yet it is difficult to agree on a precise definition. According to the Latin origins, the word means "knowledge," or better, "systematized knowledge." Science then would be a set of facts, understandings, and explanations arranged in some orderly manner. But is this definition adequate? I have knowledge that I prefer Brahms over Bartok, and I can give a dozen systematically arranged reasons why. I "know" that the mountains are beautiful any time of the year; given a year I could prove it to you. Does this "knowing" make me a scientist? Perhaps, but usually such matters of taste and judgment are excluded from science. Apparently we must specify the subject matter of science also.

To many people science is not "knowledge" at all; science is an activity. One prominent biologist insists that "science" should be a verb. "Science is investigation"; "Science is discovering new knowledge"; "Science is what scientists do." All these definitions have been suggested by college students. The last leads to some difficulties in logic when the next question, "What is a scientist?" is answered, "A person who

works in science."

Since there are at least as many opinions about what science is as there are scientists, there is no eminently satisfactory answer to the question, "What is science?" One can only be arbitrary in deciding what are the important features of science—it may even be a matter of taste. Without question, science is one of the forms of human intellectual activity. Its subject matter embraces the entire finite universe and ranges in scale from the subatomic to the cosmic. New information can be gained only by objective, verifiable observation. Mental manipulations of the observations must lead to acceptable theories, but the products of these mental activities are always subject to change if further new observations seem to demand it. Science thus strives for the most probable explanations, rather than absolute truth.

Inasmuch as science concerns itself with the whole finite universe, no one can hope to comprehend it all. We must recognize our human limitations; for this reason, science is subdivided in a number of ways. There are physical sciences, biological sciences, and behavioral or social sciences. There is pure or basic science and its partner, applied science or technology.

Biological sciences are also subdivided—on several different bases—into specialties, examples of which are given here:

1. *On the basis of the kind of organisms studied:*
   Zoology—animals   Botany—plants
   Entomology—insects   Bacteriology—bacteria
   Protozoology—protozoa   Bryology—bryophytes

2. *On the basis of the approach or the features of the organisms:*
   Taxonomy—naming and classification
   Morphology—structure and form
   Physiology—functions or processes
   Ecology—relationships to environment

3. *On the basis of the outcome of activity:*
   Basic Science—the outcome is knowledge
   Applied Science—the outcome is the solution of
       human problems

4. *On the basis of the technique of investigation:*
   Descriptive—as in taxonomy or morphology
   Experimental—as in physiology or genetics

Obviously, these lists, intended only as examples, are not complete. Even so, there is a certain amount of overlapping. An individual might be a basic scientist working by experimental methods on the physiology of bacteria. However, no professional biologist can completely lack interest in all fields of science except his specialty. In addition, certain biologists must be concerned with physical and chemical relationships. Should biophysicists and biochemists be called biologists or physical scientists?

WHAT ARE THE GOALS OF SCIENTIFIC ACTIVITY?

The public impression of the goal of science seems to be approximately the following: Science is working hard (1) to find ways of raising enough food to feed all the people, (2) to improve medicine so that the people can stay alive to eat the food, and then (3) to develop more terrible weapons to wipe out this larger, healthier population. These and other technological aspects of science make good news stories, even though they give a distorted picture of science.

Actually the goals of all scientific activities are the same: comprehension or understanding of the universe. Once an understanding of a particular phenomenon is gained, the solution of human problems may follow as a beneficial by-product. It is virtually impossible for one person to perform the whole system of activities involved in making some discovery about the universe and then to adapt this knowledge for the direct benefit of man. People specialize in one phase of activity or the other. Much of the scientific research introduced in this book is basic science, or an intellectual search for knowledge and comprehension of the living world. Ultimately, of course, all these understandings will be useful, but we leave it to others to find out just how. Fundamental biology is the foundation upon which a structure of applications useful to man is built by the applied biologists, chiefly those in medicine and agriculture.

The ultimate outcome of science would be a set of explanatory expressions or theories which would very nearly coincide with the "natural laws" that govern the universe. Once such a set of explanations is formulated, it will be easy to predict the results of any set of circumstances and then to solve any human problems. Of course, this ideal need not be anticipated in the next few years.

WHAT IS RESEARCH? This term, which can be applied to any careful study or search for knowledge, is frequently used to indicate careful library study to learn what others have thought and known before. There is a certain amount of this type of study in science because progress is a matter of building on what has been learned earlier. Usually, however, the scientist uses "research" in a slightly different way. To him, research is seeking for what was previously unknown. It involves an organized program of observation and study, resulting in new knowledge. A piece of scientific research may be conducted in a laboratory or "in the field" by one person or by a group or "team." It may be motivated by pure curiosity or by a human problem that needs a solution. In the text to follow, "research" is used almost entirely in this latter sense as a program designed to add to our knowledge facts that were not known before.

WHAT IS A SCIENTIST? If any one attribute could characterize a scientist, it would be his curiosity. I suspect that all of us have curiosity, but often it is lost somewhere in the process of growing up. A non-scientist may be content to notice an unusual event; the scientist is likely to follow through by asking, "How (or why) did this happen?"

Scientists, to be sure, have other attributes, such as reasonable intelligence, extensive training, the ability to organize ideas and information, an interest in the natural material world, and a little persistence. Fortunately for most of us, a scientist does not have to be a genius.

The personal philosophies of scientists are as individual as the scientists themselves. Each must find his own answers to a number of very serious questions. Science is supposed to be completely separated from morals and value judgments. But does this dichotomy mean that scientists can have no opinions about morals? Is it right for a scientist to seek knowledge for its own sake, feeling no obligation to point out the possible values of his findings? Should a new discovery about some aspect of cell division be considered exciting for its own sake and not because of its possible relationship to cancer? In contrast, is it possible for a scientist to feel no guilt when the results of his work are used for the purposes of destruction? Can he be excused for not speaking out on issues when his experience equips him to contribute a particular wisdom? The scientist is in a difficult position philosophically, because it is both necessary and impossible to separate the scientist as a scientist from the scientist as a human being. Since each must work out his own answers to these weighty problems, it is not surprising that there is no archetype of all scientists.

WHAT ARE THE METHODS OF SCIENCE?   Science is unique among the fields of intellectual activity in that acquisitions of knowledge occur only through objective observation. Philosophy, literature, art, all these are dependent upon the creative ability of the philosophers, writers, and artists, and they may or may not have a direct relationship to reality. Science also depends on creative mental talent, because observations can be useful only if they become part of generalizations that explain, but all of science is directly related to the real world and is dependent on this real world for its subject matter. As Santayana says, "Science contains all trustworthy knowledge." Scientific information may be discussed, interpreted, reorganized, or rejected, but in the final analysis, the only source of new information is observation.

Observation is the act (or the result) of careful, attentive watching of a natural event. Any of the senses might be used in observation, even though we commonly think of it as a result of vision. Many observations have been made with the unaided senses, but now scientific observation more commonly requires the assistance of instruments of various kinds. Magnification makes possible the observation of things too small to be seen by the eye alone. Electrical instruments allow us to observe events for which there are no senses. One of the principal aims of this book is to point out some of the methods by which we observe.

Early biology was scientific, just as modern biology is. The classical biologists observed and then formed explanations for their observations. Even today some aspects of biology are best observed in their natural state, with a minimum of interference from the observers. One pattern of observation, which might be called descriptive biology, simply studies and watches living things as they occur naturally and then describes these organisms. The pattern of distribution of a group of animals over a part of the earth's surface would be studied this way. The biologist, noting and recording most of the climatic, geological, and biological features of this area, would then be in a position to explain what factors influence the distribution of these animals. Or, as another example, the careful study, measurement, and description of a large number of individual plants could lead a botanist to an understanding of the natural kinship of various groups. Indeed, several hundred years of observations of this type have produced much of what has been known as biology.

The zoologist who studies the distribution of animals by direct observation might ask himself, "What would happen to the mice if erosion removed the topsoil from this mountainside?" He can answer his question only by waiting for this disaster to occur, or, sometimes by finding a similar mountain where it has already occurred. His observations are limited; he is helplessly dependent on the whims of nature.

In certain phases of science the experiment offers a means of avoiding this difficulty. An experiment is merely an observation in a contrived, artificial situation. Imagine how much more efficient the process of observation becomes if the scientist asks himself, "What would happen if . . . ?" and then makes the "if" happen. What would happen if a fly with red eyes were to mate with a fly with white eyes? What would happen if a plant were to grow on a soil deficient in nitrogen? What would happen if a part were separated from a living cell? Would it continue its activity? The answers to these questions can be found quickly and efficiently by experiment. The observer has some control over his observations. He can prepare a set of circumstances of his own choosing and then observe the results at his convenience.

Some aspects of biology are not easily observed by experiment, but, more and more, modern biology is using this useful observational tool which was borrowed from the physical sciences. Most of our recently acquired knowledge about the activities of living cells, most of what we know of genetics and inheritance, and most of what we have learned about the organization and coordination in plants and animals was learned by experiment.

The ideal experiment requires imagination and careful planning on the part of the investigator. The results of the experiment must be explained, of course, and the "controlled experiment" is an application of logic which makes the explanation easier. Usually the "control" is another experiment, performed at the same time, in which most but not all of the conditions are the same. Any difference in the results of the two experiments can be attributed to these differences in the conditions. The "control" thus provides a frame of reference by which the results can be judged. More detailed information on the designing and planning of experiments is included in several later chapters.

Observation is the source of new scientific information, but this information is meaningful only if it is interpreted and meshed with what is already known. The formation of explanatory statements is an equally important part of scientific activity. The results of any single observation or experiment represent a small bit of information and may be useless alone. The scientist is likely to suggest, however, that the results of the experiment are representative of a much larger body of information. If the leaves of a plant grown without nitrogen turn yellow, perhaps the leaves of all plants grown without nitrogen would turn yellow. Or, to put it another way, perhaps all yellow leaves are the result of growth without nitrogen. Either of these suggestions is a generalization founded on little evidence, but each is at least reasonable. *Inductive* reasoning has been used to arrive at an explanatory statement. Information from a particular situation has been applied to the more general case. Actually, one cannot have much confidence in these suggestions, because they result from only one experiment, and the two statements are rather different from each other. Is one of them correct? Are both correct? Each of these two suggestions must be called a *hypothesis,* a tentative explanation based on some evidence.

If the first hypothesis is correct, it should be reasonable to predict that plants raised without nitrogen, even under somewhat different conditions, should have yellow leaves. This type of reasoning is *deductive;* a general statement is used to predict what will happen in a particular set of circumstances. This new suggestion can be tested by performing another experiment. If the prediction turns out to be correct, a new, better generalization is in order, and further predictions can be made.

You might have heard that science did not develop until Francis Bacon suggested inductive reasoning. This statement is probably true, but do not forget that science uses deductive reasoning too. In fact, some philosophers argue that induction has no place in science.

SCIENTIFIC
METHOD

"Scientific method" is a term which has become distasteful to some scientists. There are many "scientific methods," and the number of steps and the order of the steps will depend upon the author of the list. Conant has presented an interesting discussion of scientific method in *Modern Science and Modern Man*. Here he gives his own description, as follows:

> *Scientists collect their facts by carefully observing what is happening. They group them and try to interpret them in the light of other facts that are already known. Then a scientist sets up a theory or picture that will explain the newly discovered facts, and finally he tests out his theory by getting more data of a similar kind and comparing them with the facts he got through earlier experiments. When his theory does not quite fit the facts, he must modify it and at the same time verify the facts by getting more data.**

In addition, he has included a "scientific method" which he obtained from a biologist:

> *Recognize that an indeterminate situation exists. This is a conflicting or obscure situation demanding inquiry. Two, state the problem in specific terms. Three, formulate a working hypothesis. Four, devise a controlled method of investigation by observation . . . or by experimentation or both. Five, gather and record the testimony or "raw data." Six, transform these raw data into a statement having meaning and significance. Seven, arrive at an assertion which appears to be warranted. If the assertion is correct, predictions may be made from it. Eight, unify the warranted assertion, if it proves to be new knowledge in science, with the body of knowledge already established.†*

If this set of steps is studied carefully, it is found to include essentially the same operations or activities described by Conant.

As I see it, scientific method is an alternation of two types of activities, the observational and the explanatory, or an alternation of inductive and deductive reasoning. Scientific method is a cyclic method of accumulating knowledge;

---

* James B. Conant, *Modern Science and Modern Man*. (New York: Columbia University Press, 1952), p. 20.
† *Ibid.* p. 20.

nowadays it never starts and never ends. Observation leads to a hypothetical explanatory induction which must be tested deductively by further observation. Each new observation produces new explanations, and each new explanation suggests new observations. At each turn of this cycle, the explanations become better; that is, they more nearly coincide with the natural laws. The hypotheses become less hypothetical as they are founded on more observations. Eventually, although it is hard to say just when, the explanations become sufficiently general, and, since the observations have been repeated by so many observers under so many conditions, widely accepted. When these statements are believed and agreed upon by the majority of competent authorities, they are often called theories, principles, or laws. Although these terms differ slightly in connotation, they are commonly used to indicate varying degrees of reliability and acceptability.

Interestingly enough, this alternation of observation and explanation works in two directions. Some individual phenomenon becomes better understood, so that more is known about this little fragment of the universe. At the same time, the observations turn up new relationships, and we find that we also know more about broader, more general aspects of the universe.

LIMITATIONS OF SCIENCE

One of the obvious limitations of science is that it can include only what can be observed. Questions of human values, ethics, morals, and religion cannot be touched by science because they cannot be observed objectively. The scientist is free to form his own conclusions on these matters as long as they do not interfere with the objective and impersonal observation in his investigations. Some scientists are atheists because they have not observed a god directly. Others call themselves agnostics; that is, they say "I do not know" because scientific methods cannot be used to investigate gods. Still others are deeply religious, believing that the order and consistency of the universe is evidence of a god. There is no inconsistency in any of these positions, and science does not force a man to believe anything that is outside of science. The only qualification is that his thinking on areas outside of science must not influence his thinking on scientific matters.

Another less obvious limitation on science depends upon the assumptions that must be made before starting any scientific activity. We may assume that something is true; we take it for granted, even if it cannot be tested by observation. There are several features of the universe that the scientist so commonly takes for granted that it may even be surprising to see them written down. These assumptions are so deeply entrenched in everyday thinking that it is difficult to conceive of their tentative nature. And yet none of these assumptions can be tested, and any of them might not be true.

We take it for granted that the universe is real. We usually do not question the reality of space, of matter, and of time. Space is there, it contains objects and matter, and things move through it. The matter can be seen and touched, and its movement takes place over a period of time. But is it possible that my whole idea of the universe is the product of the perceptions of my sense organs and of my mind? Is it possible that there is another universe, left-handed or inside-out, which occupies the same space, but which I cannot detect with my sense organs? Could time stop and with it all my mental activities, to start again only when time started again? The philosophers have debated the question, "What is reality?" We scientists have become so accustomed to thinking of our universe as matter occupying space and changing in time that it is disturbing to think that it might not be real.

While we are assuming that matter is real, we also assume that this matter is present in some amount and that it can be measured. There is some total amount of matter. Imagine the confusion if a certain piece of material varied in amount according to no predictable pattern and for no apparent cause. Measuring instruments would be useless. We would measure length with a rubber ruler.

We assume that the universe is consistent. We take it for granted that there *is* a set of natural laws, and that our scientific investigations produce theories that approximately explain these natural laws.

Another assumption is the cause and effect relationship. Every cause will bring about an effect, and every phenomenon is caused by some set of circumstances. If we understand the natural law, then if we are given a set of circumstances or a "cause," we should be able to predict the "effect."

A final assumption is that this real, consistent, deterministic universe can be comprehended by the mind of man. There is no secret in the natural laws that will not eventually be explained if enough observations are made. There is no great mystery, no whimsical maker and changer of laws, that is beyond the mental ability of man. This is a very necessary assumption, as without it all investigation is futile. If we cannot understand the universe anyway, and if there is no hope of solving our problems, why bother to try? The defeatist attitude is a natural consequence of the failure to make this assumption.

These assumptions are so much a part of our thinking that we are hardly aware that we take them for granted and that they cannot be tested. It is almost beyond reason to admit that any one of these might not truly represent the universe. Many of the best science-fiction stories depend upon the failure of one or more of these assumptions.

Yet, at least one of them is a little shaky. The cause-effect relationship works on the ordinary size level but must be slightly modified on the atomic or the cosmic level. Instead of effect from some cause, it is possible to give only a "most probable effect." Individual atoms and subatomic particles do not necessarily obey the law of cause and effect. The new theoretical physics, wave mechanics or statistical mechanics, has not yet exerted its full influence on science. It will be interesting to watch the revolution in thinking when it does.

SELECTED REFERENCES

Conant, James B., *Modern Science and Modern Man.* New York: Columbia University Press, 1952. This is one of a series of short books in which Conant discusses science and particularly the understanding of science by non-scientists.

Lachman, Sheldon J., *The Foundations of Science.* Detroit: Hamilton Press, 1956. Contains a clear description of science, its activities, goals, and conceptions. The writing is clear and easy to follow. In some cases it is deceptively simple; some of the paragraphs are more profound than they seem at first reading.

Polanyi, Michael, *The Study of Man.* London: Routledge and Kegan Paul, 1959. *The Study of Man* is a condensation and extension of ideas expressed in the longer book, *Personal Knowledge.* Polanyi is a chemist who has become concerned with how we know things. One may not agree with him, but he certainly stimulates thinking.

Tullock, Gordon, *The Organization of Inquiry.* Durham, N.C.: Duke University Press, 1966.

It would be very easy to suggest that a major revolution has occurred in biology since the end of World War II; certainly there are more biologists using physical and chemical methods. Much of the recent progress in biology can be traced to a shift in emphasis from descriptive to experimental biology, but to call this a major revolution would probably be going too far.

Of the important factors in the encouragement of experimental research, one has been the increased availability of funds from several sources. Certain instruments and techniques—such as isotopic tracers, optical devices, and chromatography—have become generally available, but none of these was really new in 1946. When money became available, more people could work in experimental laboratories, and the laboratories could be better equipped. Several important instruments could be placed in production economically. The modern experimental biology laboratory contains an array of sparkling expensive instruments, a fair share of which would necessarily have been homemade only a generation or so ago.

In addition to a better financial situation, there is now a rather different body of concepts upon which to build. The discovery that nucleic acids can control the synthesis of enzymes has completely altered the course of both genetics and enzyme biochemistry. The knowledge that living cells can be broken and that the parts will continue some of their activities even when separated from the rest of the cell has had a profound influence in cellular physiology.

Human beings have been biologically minded for a very long time, since man could never have become civilized without being aware of the world of living things. Biology has been scientific in the usual sense, however, for only a few centuries. For a number of reasons biology has always developed only after developments in the physical sciences.

The earliest scientific biologists were descriptive scientists, concerned with naming and describing organisms. The invention of the microscope permitted the examination of smaller units, but the approach of the biologists remained about the same; that is, they still described what they saw. In time, it became possible to consider more abstract relationships. Changes that take place over a period of time, as in the growth of an animal or plant, or as in the various physiological processes within an individual, obviously require a

12

rather different kind of observation. If a biologist is to explain
the interrelationships of the activities of various organs in
an animal, a higher level of intellectual activity is required
than if he merely describes their structure.

The experimental approach is especially valuable in these
highly abstract phases of biology. In fact, about 1625 William
Harvey used some of the first experiments in biology to
demonstrate the continuous circulation of the blood. At about
the same time, van Helmont's experiment showed that plants
do not take all their food from the soil. About a century later
Stephen Hales performed many experiments on the pressures
and the movement of liquids within plants and animals. His
books, *Vegetable Staticks* (1727) and *Haemastaticks* (1732),
demonstrate the ingenuity of the man, as well as providing
very interesting reading. Physicists at the time were measuring
pressure by observing the height to which a liquid would rise
in a tube. Hales applied similar observations in biology. In
one heroic experiment he measured the blood pressure of a
horse by attaching a long glass tube to one of the large arteries
in the horse's neck. Fortunately this method of measuring
blood pressure never became popular among the physicians.
It is enlightening to read a modern laboratory manual for
plant physiology, to compare it with *Vegetable Staticks,* and
to note how many of the usual experiments in present-day
courses were actually designed by Stephen Hales.

One of the major philosophical battles in the field of
biology was the mechanist-vitalist controversy about the end
of the nineteenth century. The mechanists maintain that there
is nothing about the living organism that cannot be explained
in terms of physics, chemistry, and mathematics. According
to the vitalists, there was some vital force, some living being,
above and beyond the physical laws. Progress in the biological
sciences has depended on the assumption of the mechanistic
interpretation. We take it for granted that life can be explained
by physics and chemistry and that living things obey the
physical laws. This assumption, although it cannot be tested,
is very necessary because the vitalist assumption acknowledges
that life cannot be understood by science. Even if we think
and experiment as mechanists, a number of serious problems
still exist. The goal of all biological research must be the
explanation of life according to the physical laws. Therefore
an understanding of living things can never be more complete
than our comprehension of the physical laws. Biology must

always lag behind the physical sciences, just as physics has lagged behind mathematics. Allowing investigations in biology to contribute to the physical theories is the only alternative.

DIFFICULTIES IN BIOLOGY

Biology offers several other challenging difficulties. The living system is far more complex than any physical entity. Biological chemicals are large molecules, and the arrangement of these molecules in a particular pattern is a necessary condition for life. No mere list of all the kinds of chemical compounds and the amounts of each will ever explain life, and yet, no picture showing the exact location of each molecule within a cell would be adequate either, because these locations change with time. Altogether, the living cell is more complex than anything the physical scientist is used to working with. In fact, some modern scientists feel that the whole living organism is more than the sum of its parts. Although reminiscent of vitalism, this new approach is perhaps on better intellectual grounds and even suggests that findings in biology can contribute to understanding of the chemical and physical laws. The little book by E. Schrödinger, the great theoretical physicist, called *What is Life?,* gives an extremely stimulating discussion of this idea. Even though the ideas were first proposed in 1943, they are still pertinent.

Living things are more difficult to observe than nonliving things. Their extreme complexity is only a part of the difficulty. Living material is naturally variable. Its responses during an experiment may be influenced by a great variety of factors, and the experimenter may not be aware of some of these. Often a difference in the treatment of an organism before the experiment will cause changes in responses. Some chemical compounds can have profound effects even if the organism contains only a few molecules per cell.

Ideally, the experimenter hopes that the experimental treatment will not influence the results, but this desire can never be realized. The best that can be hoped for is to reduce the interference to a minimum. Obviously, if the techniques of the experiment influence the organism, controlled experiments are essential before any conclusions can be drawn.

THE BIOLOGIST'S ASSUMPTIONS

Although some of them have been mentioned previously, it is probably wise to explain the assumptions which, in addition to the general assumptions of science, are made by the biologist.

1. Living organisms obey the laws of physics and chemistry. This item has already been discussed. The only step necessary to prevent the invalidation of this assumption is to incorporate any inconsistencies discovered in biology into the physical laws.

2. The whole living organism is nothing more than the sum of its parts. This assumption is derived from the analytical technique used by physicists. They "analyze" complex systems into manageable subsystems, and then study each subsystem separately. Once the subsystems are understood, the separate theories can be "synthesized" into a whole. Many biologists are reluctant to apply this approach to organisms, the first argument being that living things are highly coordinated. It would not seem unreasonable, however, to assign this very coordination to a separate subsystem. Within the middle ground between biology and physics great debates are now being conducted between those who approach the area from the physics side and those from biology. Part of the problem is probably semantic, in that the two parties do not understand each other's language. However, we certainly cannot belittle the fact that physical approaches have made great contributions to our understanding of organisms in the last twenty or thirty years.

3. The experimental treatment does not affect the process being investigated *too* much. This is the hardest of the assumptions to make with any confidence. The biologist often must use considerable ingenuity in the planning of experiments in order to feel confident in interpreting the results.

4. Related organisms, or parts thereof, will have identical or similar behavior under the same circumstances. This assumption is not always needed, but is especially useful when dealing with human problems. Most experiments with human beings would be both unethical and immoral, so we must depend upon information from other animals.

BIOLOGICAL PROBLEMS   Many of the most serious problems remaining for the biologist today are problems which can only be solved by experimental methods. Even the areas of biology which have classically used descriptive methods are now turning to experiments. The taxonomist who used to be concerned only with the sizes, shapes, and colors of various parts now performs experiments to learn the effect of environmental changes on these sizes, shapes, and colors. The embryologist who formerly described the embryo at the ages of one, two, three, . . . n days has become concerned with the reasons for the observed changes.

The general nature of the major remaining problems in biology might be grouped into the following interrelated categories:

1. Relationships of materials
2. Energy relationships
3. Control and integration phenomena

The relationships of materials, that is, the kinds of chemical compounds present and the chemical reactions that occur, make up the province of biochemistry. The general pattern of cellular biochemistry has been evolved within a period of about the last thirty years. Many details of cellular chemistry remain to be discovered, but as most of the techniques are available, it is difficult to visualize any major changes in concepts. Such statements are dangerous, however, because new concepts have a way of appearing without advance warning.

Energy relationships involve a more difficult problem. Energy is an abstract concept in its simplest forms. Living organisms bring about transformations of chemical energy, heat, electricity, motion, and light, from one to almost any of the others, usually with high efficiency. The physics comprising these energy relationships is among the most interesting and challenging of biological problems.

The control and integrative systems constitute a series of almost purely biological problems. Certainly some man-made control systems operate on similar principles, but it might not stretch the imagination too much to call these systems biological, at least in origin. The questions of the regulation of all of the various processes within an animal, or the integration of the activities of all the individual plants or animals within a group, are among the most interesting but most frustrating. It has been exceedingly difficult to find means of investigating these elusive phenomena.

Any attempt at classification of problems or areas of inquiry is certain to lead to oversimplification. A more realistic picture of the questions which still face the biologist can be gained by examining some problems as examples. In most of these, all three of the categories above are involved.

What controls the development of an organism? How is it possible that the fertilized egg of a chicken always develops into a chicken and never into a duck? How are the transformations of groups of cells into tissues and organs controlled? The nucleic acid is the hereditary material, but how does a nucleic acid molecule bring about such profound changes and differences? If nucleic acids control the synthesis of enzymes, what determines when a given enzyme is formed and when it is not? If a system of regulator genes controls the timing of enzyme synthesis, how could such a system have evolved?

How can the cell membrane exhibit such control over what enters and leaves the cell? How is energy utilized to move molecules or ions in what seems to be the wrong direction, that is, against diffusion gradients? How does a kidney cell sort molecules and ions? How can plant cells continue to absorb mineral ions even after they are more concentrated inside the cells than in the surrounding soil or water?

How do some cells or organisms measure temperature and time? How can the hypothalamus control temperature so precisely in mammals and birds? How does it measure temperature and what does it use as a reference point? How do plants and animals measure the length of a day or night in order to respond to the seasons? If all the birds in a flock suddenly change their direction of flight, is it because they have communicated with each other or because all have detected the same minute change in environment? If all the flowers in a field suddenly burst into bloom, is it because they have communicated with each other or because all have detected the same minute change in environment?

How do cells convert light energy into useful chemical energy as they do in photosynthesis and in vision? What chemical and physical processes are involved in color vision? What can be said about the molecular biology of the brain during the storage of information in the memory?

What is the nature of the specific structure of the minute parts of cells? How are enzymes and other molecules arranged in space to allow sequences of chemical reactions to proceed in an orderly fashion? How can the minute structures seen in electron microscope pictures be related to the observed chemical changes in cells?

Such a list of questions makes it seem that the biologist does not know very much. The magnitude of some of the questions should make anyone humble, but at least we now know some of the questions. In their attempts to answer these and other questions, biologists have used a great variety of approaches. Some have led to the great progress we have already made; some have ended in frustration. However, biologists are a persistent lot. They continue their attempts with whatever techniques are available. Certainly the attempt is worth the effort, because an understanding of these problems will be a truly noble achievement.

SELECTION OF A RESEARCH PROBLEM With all the serious questions that remain to be answered in biology, why does the beginner in research frequently have a difficult time picking a problem to work on? This seems especially to be true if he is working alone. Ideally the beginner starts under the supervision and careful guidance of an experienced master. Even so, a few general rules are probably in order, even if it is easier to agree with them than to obey them.

1. Experimental research should be motivated by curiosity, the desire to know. There are other reasons for doing research, of course, ranging from a good grade in a course all the way to advancing one's professional career. Nonetheless, if these reasons are not secondary to the basic curiosity, the quality of the research will suffer.

2. The investigator must already know a great deal about the subject he chooses to investigate. More often than not, the problems studied in research laboratories have arisen from other problems. If the beginner has had no experience, he must depend upon the experience of others by reading original technical papers in the journals. Every student finds that some topics appeal to him more than others and within the area of interest there will surely be a large body of literature. As he reads, he soon finds himself thinking, "Why did they do this?" or "Why didn't they do the next logical experiment?" Frequently, also, observations made in instructional laboratories will initiate rather similar questions. In either case, further reading and discussion will focus the attention on a problem area that almost demands experimental satisfaction.

3. A useful point to remember is that some problems promise results more surely than others. Too frequently students recognize as potential research areas only those problems that have received publicity because they have challenged the very best biologists. One year all the students

wish to cure cancer; another year they want to study DNA chemistry; the next year, organ transplants. The beginning researcher is unlikely to contribute much in such a problem area. The student must realize that his facilities and time are limited. At the end of the time available he may be asked to write a paper describing his results. But if all his experiments were failures, what does he write about? If the problem is selected judiciously, the results may be meaningful no matter which way the experiments turn out.

SELECTED REFERENCES

Beveridge, W. I. B., *The Art of Scientific Investigation.* Revised edition. New York: W. W. Norton & Co., 1957. This is a friendly and personal discussion, particularly related to the mental processes, creative activities, and responsibilities of the scientist. It is delightful reading, as it is sprinkled with anecdotes from current and historical biology.

Bonner, James F., *The Molecular Biology of Development.* New York: Oxford University Press, 1965. A report of the work, primarily from one laboratory, on one of the challenging modern areas of biology, including problems, experimental approaches, and interpretations.

Commoner, Barry. 1961. In defense of biology. *Science* 133:1745–1748. (Reprinted in Blackburn, ed., *Interrelations: The Biological and Physical Sciences.*) A pointed argument, attempting to put "life" back in biology.

Gabriel, M. L., and S. Fogel, eds., *Great Experiments in Biology.* Englewood Cliffs, N. J.: Prentice-Hall, Inc., 1955. A collection of papers from the original biological literature, reprinted from the periodicals in which they appeared. Each paper has been selected to demonstrate some quite significant finding. The collection as a whole is a group of specific examples of biological research.

Gerard, R. W., and R. B. Stevens, eds., *Concepts of Biology.* Washington, D. C.: National Academy of Science-National Research Council Publication 560, 1958. This is an transcription of several days of discussions held by a group of eminent biologists, including both conservatives and liberals. Although they did not reach complete agreement on what should be the major concepts of biology, the discussions cover a great range of ideas.

Schrödinger, Erwin, *What Is Life?* New York: The Macmillan Company, 1945. (Pertinent parts reprinted in Blackburn, ed., *Interrelations: The Biological and Physical Sciences.*) Every biologist should read this discussion of the relationship between the biological and physical sciences, even though it is several years old. The findings of some of the most active years of biological research do not detract from the stimulating discussion. The general theme is contained in the following quotation from pages 68 and 69: "Living matter, while not eluding the 'laws of physics' as established up to date, is likely to involve 'other laws of physics' hitherto unknown, which, however, once they have been revealed, will form just as integral a part of this science as the former."

Twitty, V. C., *Of Scientists and Salamanders.* San Francisco: W. H. Freeman and Company, 1966. A delightfully written and highly instructive account of a lifetime spent in experimental zoology.

Watson, James D., *The Double Helix.* New York: Atheneum Publishers, 1968. A Nobel laureate's story—highly personal, somewhat controversial, often entertaining—of the formulation of the now famous model for DNA.

THE BIOLOGICAL LITERATURE

Progress in science builds upon previous progress in science, which, of course, is possible only because written records have been maintained for centuries. Scientific achievement depends only partly on brilliant experimental work and astute observation; the formulation of useful explanations depends upon the interlacing of current work with the observations and interpretations of the past.

Since a research program becomes meaningful only if it is interrelated with all the previous work on the subject, the need for maintaining permanent records is obvious. The "biological literature" comprises all that has been written on biological subjects in the last several hundred years. All this writing can be subdivided into two general types: (1) primary publication describing the original observations, and (2) secondary publication which summarizes, describes, or discusses these first reports. Almost all the primary reports now appear in periodical or serial publications, or "journals." Secondary discussions and summaries are considerably more varied.

LABORATORY RECORDS

A part of biological literature not generally accessible includes original laboratory records and notebooks. Everyone who performs experiments in the laboratory has an obligation to keep records of what was done and what happened.

The form in which these laboratory records are kept is a matter of personal preference. Some people like to write up each experiment individually on separate sheets of paper—describing the materials used, the methods, and the results—and then file this unit. Probably a more satisfactory procedure is to write everything into a bound notebook. A description of the reason for doing the experiment, the minute details of materials and methods, complete notes on the results (including even observations that may seen inconsequential), and finally, calculations and discussions can all be included in a few succeeding pages of the notebook. Bound notebooks containing quadrille-ruled paper, that is, paper ruled in squares, are useful because preliminary graphs can be recorded directly with the results.

Regardless of their form, these original laboratory records are the most important part of the permanent records because it is upon these that the rest of the literature is built.

TECHNICAL
PAPERS
Once a piece of experimental research has reached a rea-sonable conclusion, the work is described in a technical "paper." Generally this is a brief report, describing the work in as much detail as ever will appear in the public literature. The organization of the report varies somewhat, but the paper usually includes a statement of the problem, a descrip-tion of experimental methods, a summary of results, a section on interpretation or conclusions, and a list of references to cited literature.

Many serial or periodical publications are devoted almost exclusively to these technical papers. Some, like the *American Journal of Botany* and the *Journal of the American Chemical Society,* cover an extremely wide range of subjects. Others are restricted to a single group of organisms or even to a single technique. Table 3-1 lists as examples a number of journals which are valuable to the experimental biologists. Some of these are sponsored by organizations, societies, or

TABLE 3-1   *examples of periodicals useful to the experimental biologist*

technical journals	review periodicals
American Journal of Botany	*American Scientist*
Archives of Biochemistry and Biophysics	*Annual Reviews of:*
	Biochemistry
Biochemical Journal	*Microbiology*
Biochimica et Biophysica Acta	*Physical Chemistry*
Doklady Akad. Nauk S.S.S.R.	*Plant Physiology*
Journal of the American Chemical Society	*Bacteriological Reviews*
	Biological Reviews of the
Journal of Biological Chemistry	*Cambridge Philosophical*
Journal of Cellular and Comparative Physiology	*Society*
	Botanical Reviews
Journal of Molecular Biology	*Physiological Reviews*
Plant Physiology	*Scientific American*
Zeitschrift für Naturforschung	

institutions, and others are produced by commercial pub-lishing firms.

Since the primary technical report is the most detailed form in which a piece of work will ever be described in the literature, it is imperative that this description be adequate. The report should permit evaluation of the work and should allow repetition by persons in other laboratories. If the length of the report had no limitations it would be easy to include more details than necessary, to take no chances of leaving out something essential. Unfortunately, papers must be short; there simply is not room in the journals to publish every paper at full length. The result, of course, is a compromise in which the writer assumes that the reader is familiar with ordinary laboratory techniques. The technical paper, then, is a report in which one research worker communicates with others like himself.

A technical paper could be written in almost any language, but now most are in English, German, French, or Russian. A little of the literature from Japan is in the Japanese language, but fortunately for us, most of the Japanese workers now write in English. Papers from the Scandinavian countries are likely to be written in English or in German. Most American biologists can read at least one other language, but almost none could write a paper in any language except English. Some Americans unrealistically solve the language problem by simply ignoring the foreign literature.

REVIEWS   So many technical papers appear each year that no one could hope to follow the whole literature. A valuable aid is the review article, which attempts to summarize and evaluate some small segment of the literature. For example, any one year might see the publication of several hundred technical papers all somehow related to muscle contraction. The author of a review article on muscle contraction examines all available papers on the subject in great depth and prepares a summary of current thinking in the field as he sees it. The review article thus helps to pull the literature together, but it is subject to the opinions and judgment of the author. This restriction is not altogether bad because where there are two schools of thought a member of the other school is likely to write the next review article on the subject.

The review paper typically contains no new data. Instead, selected tables or graphs from the previously published reports may be reproduced. The review article might cover one small area completely, reviewing all the work ever done on the subject. More commonly, each review paper covers the work of some brief recent period, such as the previous year. The writing and publication of the review take time, so the review will be one to three years behind the original work. A large number of references will be listed to enable the reader to consult the original papers for details.

Papers in this category may be at any level of technicality. Several examples of review publications are given in Table 3-1. Some are highly technical; others are written for the general scientific or educated public. *Physiological Reviews,* for example, is written for physiologists, but the review articles in *Scientific American* are aimed at the educated public. Many of the *Scientific American* articles succeed admirably.

MONOGRAPHS, SYMPOSIA, BOOKS

A monograph or book describing the work of one individual on one subject used to be the only way of publishing scientific information. Any monograph published today is usually a long review, particularly covering one subject or the work of one laboratory. For example, taxonomic biology produces monographs describing one genus or one family of organisms.

One of the relatively recent innovations in the scientific literature is the symposium volume. Increased availability of funds and modern rapid transportation have made it possible for institutions or societies to call together the various persons working on some subject to discuss recent work and problems. This meeting, or symposium, may be highly formal or very loosely organized, but generally each contributor summarizes the work of his laboratory or comments at length on interpretations. The "minutes" of such meetings, whether as formal reports or as transcriptions of informal discussions, can be exceedingly valuable.

ABSTRACT SERVICES

Since no one can hope to read every issue of every periodical, some system for keeping up with the literature becomes essential. The various abstract services perform this function quite well.

An abstract is a one-paragraph summary of a technical paper. It may be written by the author of the paper or by some other scientist familiar with the field who uses the technical paper as his source of information. So that any reader can find the original paper if he chooses, the abstract lists the title, authors, the periodical with volume and page numbers, and the date of publication. Ideally the abstract should include all the information contained in the original paper, but to compress several or many pages into one paragraph without loss of information would take extremely careful writing. This ideal is approached but never attained.

The biologist depends upon *Biological Abstracts,* a periodical which presents abstracts of a very large segment of the world's new biological literature. Two issues appear each month, and each issue is subdivided into general topics. Another of the world's leading abstract services is *Chemical Abstracts,* published semimonthly by the American Chemical Society. Since much of the modern work in biology includes some chemistry, *Chemical Abstracts* is as useful as *Biological Abstracts.* Both cover many of the foreign periodicals which are not available in our libraries.

Computer techniques and modern filing and communications procedures make it possible for *Biological Abstracts* and *Chemical Abstracts* to describe a paper in surprisingly short time after its original publication. The same computers prepare lists of titles of papers alphabetized according to key words. Figure 3-1 shows an example of such a list. Both *Biological Abstracts* and *Chemical Abstracts* now publish such computer-prepared lists. The investigator can look up the subject words he is interested in and quickly find the abstracts that are important to him.

Current Contents is useful also, even though it contains no abstracts. This periodical, which was started only a few years ago, reprints the tables of contents from a large number of journals. The papers appearing in the current issues can be surveyed more quickly than by looking up the journals themselves.

OTHER LITERATURE   A few periodicals do not seem to fit into any special category. *Science,* published by the American Association for the Advancement of Science, is such a general periodical. It contains some original technical reports but also two or three reviews in each weekly issue, along with editorials, news of science, news about scientists, book reviews, and advertisements. *Nature* is a British publication similar in coverage.

FOLLOWING INJECTION OF	PICRO TOXIN INTO THE LATERAL CEREBRA	2177
AL INFARCTION/ CLINICAL	PICTURE AND DIAGNOSIS OF THE INTER V	1257
CHARACTERISTICS, BLOOD	PICTURE AND INDICES OF THE INNER ORG	824
D GOAT IN MAN/ CLINICAL	PICTURE AND THE TREATMENT OF SEQUELA	2829
DRENAL-CORTEX/ CLINICAL	PICTURE AND THERAPY OF CONGENITAL VI	1542
THOGENESIS AND CLINICAL	PICTURE OF ACUTE DISORDERS OF CEREBR	1311
H AND YOU. A GEOGRAPHIC	PICTURE OF THE WORLD WE LIVE IN/ THE	34
TIGATION OF THE PEABODY	PICTURE VOCABULARY TEST WITH THREE E	1898
EA-PIGS IN HISTOLOGICAL	PICTURE/ GENERAL CHANGES OF ORGANISM	2735
SIVE AND NON-AGGRESSIVE	PICTURES/ THE EFFECT OF AN UNDETECTE	1892
ANATOMY AND ECOLOGY OF	PIERIS-PHILLYREIFOLIA-□HOOK.□-DC/	3204
MIGRATION OF	PIG EMBRYOS FOLLOWING EGG TRANSFER/	2441
PLACE OF IN BREEDING IN	PIG FARMING/ THE	2497
NATION WITH ADEVETIN IN	PIG REARING/ THE RESULTS OF EXPERIME	2496
SWINE FETUS AND NEWBORN	PIG/ INTESTINAL LACT ASE, ALKALINE A	1099
BIOCHEMICAL CHANGES IN	PIGEON BREAST MUSCLE MITOCHONDRIA FO	1629
STIMULUS DIFFERENCE□ IN	PIGEONS□/ DISCRIMINATION AND GENERAL	363
S IN THIAMINE DEFICIENT	PIGEONS/ ENZYME STUDIE	964
FAMILIAL HYPER	PIGMENTATION OF THE EYELIDS/	269
THE PHYSIOLOGIC ROLE OF	PIGMENTATION. AN EXPERIMENTAL STUDY/	1705
LOCOCCAL LIPO LYSIS AND	PIGMENTATION/ STAPHY	2662
FAMILIAL INCONTINENTIA	PIGMENTI/	1681
PARATION OF CHLOROPLAST	PIGMENTS BY COUNTERCURRENT DISTRIBUT	3210
D INHERITANCE OF FLOWER	PIGMENTS IN DIPLOID MEDICAGO SPECIES	203
CHS/ THE VISUAL	PIGMENTS OF SOME DEEP SEA ELASMOBRAN	1725
OMBOPLASTIN FORMED FROM	PIGS PLASMA IN THE THROMBOPLASTIN GE	1319
DURING THE CLOTTING OF	PIGS PLASMA/ THE NATURE OF THE THROM	1318
THE EARLY WEANING OF	PIGS/	2491
THE SMALL INTESTINE OF	PIGS/ GLUTAMIC OXALOACETIC TRANSAMIN	1078
AGE AND PRODUCTIVITY IN	PIGS/ RELATIONSHIP BETWEEN	1502
REPRODUCTIVE ORGANS OF	PIGS/ THE BLOOD SUPPLY OF THE	1195
LENGTHS. EXPERIMENTS IN	PIGS/ THE ELECTROLYTE AND PROTEIN CO	1079
HISTORY OF THE NORTHERN	PIKE, ESOX-LUCIUS-L□ PISCES□, WITH S	489
ASEXUAL REPRODUCTION OF	PILOBOLUS/ LIGHT AND THE	3146
LCHOLINE, HISTAMINE AND	PILOCARPINE/ COMPARATIVE SENSITIVITY	2008
THE	PILOT HOME CARE PROGRAM OF TORONTO/	2927

FIGURE 3-1   *Part of a page from a computer-prepared list of key words. Numbers following titles identify the abstracts in the accompanying issue of Biological Abstracts.*

Letters to editors have become an important means of communication, many periodicals reserving a section of each issue for such letters. The usual technical paper requires a fairly long time for editing and publication and may not appear in print for six months or a year after the manuscript is submitted to the editor. If for any of several reasons an author wishes to have information published much faster, he can write a letter to the editor. This letter is not subjected to the same editorial scrutiny as the formal paper, so the new data or interpretive discussions can be printed earlier than if a technical paper were written. Presumably, the letter to the editor will be followed at some later date by the complete technical report. Unfortunately some people have a tendency to publish all their work as letters to editors, and they never get around to publishing the full details. This practice is not acceptable, of course, but it is very difficult to regulate.

Some other forms of the literature useful as sources of information are handbooks of various sorts. The physical and chemical handbooks give tabulated data on almost any imaginable subject. The FASEB Handbooks include specifically biological information. The chances are good that if a fact about an organism is known, it has been included.

A source of ideas that is sometimes neglected is the collection of materials provided by the concerns that make or sell supplies and equipment. The catalogue of a glassware manufacturer, for example, might contain sketches of specialized pieces designed for petroleum testing that could be used for biological experiments. The exchange of ideas among the various sciences as a result of reading catalogues and advertisements is probably more important than many people realize.

The successful biologist depends upon a variety of sources of information. One of his most valuable tools is the kind of library provided by a university, and one of the most valuable techniques he can learn is to use the library effectively.

SEARCHING THE LITERATURE Anyone working in a scientific field must keep himself aware of other work in that field. The scientist just entering a new area has a particularly difficult task because the whole literature is unfamiliar to him. Anyone who does not bother to read the previous literature, but instead goes into the laboratory and starts experimenting, faces the very real likelihood of an unnecessary duplication of effort. He must realize that his mental processes are not unique and that someone else has had or will have ideas similar to his own.

No one can hope to read all the literature, even in a fairly small segment of science. In fact, new literature is appearing more rapidly than one individual can list it, let alone read it. *Biological Abstracts* is now presenting abstracts of almost 100,000 papers each year. *Chemical Abstracts* provides coverage of about 7000 journals. Even reading these two abstract periodicals from cover to cover is impossible or at least is a full time job.

A new field can best be approached by first getting a broad general picture. One simple means of accomplishing this objective is to learn it first-hand from someone familiar with the field. If this is impossible, a recent textbook provides a point of beginning. The textbook may be a dead-end source, but most now include at least a few references to more advanced literature. Several encyclopedias also are excellent starting points. Encyclopaedia Britannica and Encylcopedia Americana give quite detailed coverage of many biological topics, and many of the articles provide references. The McGraw-Hill *Encyclopedia of Science* is generally excellent. Several one-volume specialized encyclopedias have appeared recently, such as the *Encyclopedia of Microscopy* and the *Encyclopedia of Spectroscopy.*

Eventually, references obtained from textbooks and from encyclopedias take one to the review articles, and references given in these lead quite directly to the original technical reports. By this time the reader will have a good idea of what is known in the specific field. Other papers can be found in the abstract journals. *Chemical Abstracts,* for example, publishes annually a subject index, an author index, and an index of chemical formulae. At ten-year intervals, they produce decennial indexes. If you wished to find all the papers on a subject covered by this service, you could trace the annual indexes back to the most recent ten-year index. The abstracts you find will tell you whether it is worth seeking the original paper.

A *systematic* literature search would probably include the following steps, although not necessarily in this exact order:

1. Gain an overview of the subject from appropriate textbooks.

2. Develop a card system for recording the articles you find.

3. If the textbooks have provided references to recent review articles that sound promising, look up and read those articles. Note any cited references that seem to apply to your field of interest.

4. Follow up those papers in the original journals, again noting any pertinent references. You have started at the base of a "tree" with branches that become narrower as they spread more widely.

5. Go to the subject matter indexes of *Biological Abstracts, Chemical Abstracts*, or both if appropriate. If there are other appropriate specialized abstracts, such as *Helminthological Abstracts,* consult those also. *Dissertation Abstracts,* as the name implies, covers papers presented by candidates for advanced degrees. These papers may not be published in the regular journals.

6. By this stage you will have discarded many papers that seemed at first to pertain to your subject. You will also

find that papers you need will come to your attention from more than one source. You can then really begin to systematize your records.

7. Pursue the subject backward in time to its origin. If it is necessary to go back to times before the abstract services were begun, you may need to consult the original journals directly. Fortunately, at that time there were not so many journals as today, and you will know which ones are likely to carry papers on your subject.

Somewhere you must stop your search through literature, lest it go on forever and keep you out of the laboratory. Just when to consider your knowledge of the field adequate is a matter of personal judgment. I think we learn primarily by experience.

REPRINTS When a scientific periodical is printed, extra copies of each paper are usually run off. These are not bound with other papers, but are given or sold to the authors. These separate copies, or "reprints," are printed from the same plates, and therefore are identical to the published paper. The author distributes these reprints to persons he thinks would be interested or to other scientists who request them. Anyone working on cellular metabolism could request reprints from other writers on the subject and thus could have his own collection of papers for careful study. In turn, he provides reprints of his own papers to the other workers. Sometimes it is easier to obtain a reprint from an author than to obtain the journal in which the paper was published.

RECORD SYSTEMS Since the literature is voluminous—every year we read or become aware of a vast number of papers pertaining to our work—some system of recording these papers is essential. Most such record systems consist of filing cards of one sort or another.

Each researcher must develop his own system for keeping track of the literature. For cards to be useful, there should be at least one card for each paper, and that card should give authors; title; citation to journal, volume, year, and pages; and preferably a brief summary of contents. The information on the cards is useless unless it can be recovered when needed.

It must not be thought that some system is effortless; any system requires expense and work. Elaborate cross-references become necessary if the file of cards is large. You might choose to file the cards alphabetically by author, but many papers have more than one author, and the best-known author is not always listed first.

The system described in the following paragraphs is included as an example only and cannot be recommended without qualification. It is doubtless as expensive and difficult to keep up as any other system, but it does offer the advantage of rapid retrieval of any card in the file.

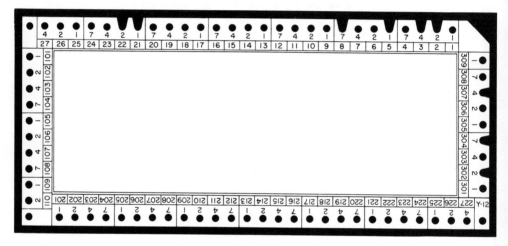

FIGURE 3-2   *One form of edge-punched card.*

Several firms produce edge-punched cards, of which Fig. 3-2, is an example. Many sizes are available, some of which have several rows of holes. The card shown is about the size of an IBM card, for which filing cabinets are easily available. The holes can be punched out, as several in the illustration are, so that when a needle is passed through a given hole in a stack of cards only those that are punched will fall off the needle. One corner is cut off the card so that the cards in a stack can be quickly sorted and placed in order, all right-side-up and facing forward. Each card is prepared by copying the authors' names, the title of the paper, citation, and an abstract on the face of the card. The holes in the card are then punched according to a code.

Holes are arranged in "fields" of four, labeled 1, 2, 4, 7.

By punching no more than two holes it is possible to arrive at any digit from 0 to 9, as 1, 2, 2 + 1, 4, 4 + 1, etc. If two fields are used, one field can be used for units, the second for tens. For example, 7 + 1 in the first field is 80, 4 + 2 in the second field is 6, and the two together give 86. The card illustrated has 19 fields, which means that 10^{19} coded bits of information can be stored. This is not fully efficient because a set of four holes could stand for any number from 0 to 15 if the number of punches per field were not limited to two. Therefore the card could store 2^{76} (about 10^{23}) bits of information.

The type of information stored can be described briefly. Authors' names are given a code designation, a number from 00 to 99, by dividing the alphabet into 100 more-or-less equal categories and using two fields for the name of the first author on the paper. A second pair of fields gives the second or best-known author. The year of publication requires one pair of fields plus one extra hole for a few papers before 1900. Journals have been coded by a modified alphabetical scheme. If a straight numerical alphabetical code were used, some categories such as BIO and JOU would be overloaded; for this reason certain journals are given a separate numerical designation.

Since most papers can be classified in more than one way, subject matter coding uses two sets of three fields each. A decimal system similar to that used in the library has been adopted. The subject matter file is the most difficult to work out. It is necessary to decide which are the broad general categories and then to decide what are the reasonable subdivisions of each. The subject matter field must be reasonably detailed and yet must be flexible enough to be modified when your ideas of what is important change, as they will in the future.

The preparation of the cards requires a great deal of time, of course, but then so does the use of any other card system. Once the cards have been prepared, there is no need to file them in any particular order. From among, say, two thousand cards, all the papers by one author can be found with four or five passes of the needle, followed by hand sorting of an average of twenty cards in that alphabetical category.

I reiterate, this system can work, but everyone must develop his own scheme of coding. Storing and sorting by computer would be even easier, but a good deal more expensive. Sorting by hand from a set of cards arranged in alphabetical order might be faster, but the file must be kept in alphabetical order and cards must be duplicated for cross-reference.

SELECTED Most of the reference works mentioned in this chapter are
REFERENCES so generally useful that they are listed in the Bibliography
at the end of the book.

Casey, Robert S., James W. Perry, Madeline M. Berry, and Allen Kent, *Punched Cards, their Applications to Science and Industry,* 2nd ed. New York: Reinhold Publishing Corporation, 1958. A description of the cards and equipment available, descriptions of applications to research (including biology), and valuable suggestions on coding.

Bryan, John H. D. 1966. Information retrieval system based on edge-notched cards. *Bioscience* 16:402–407.

Classical or descriptive biology has developed a tremendous vocabulary for describing various aspects of organisms. Even a description of an organ of a plant or animal, however, must eventually depend upon some quantitative measurement. The fact that the tip of a leaf is more-or-less rounded is frequently less important than some estimate of the overall size of the leaf.

Experimentation in biology certainly could not function except in a quantitative manner. If materials are produced by cells they are produced in some quantity. If a part of an animal moves, it moves some distance. If some environmental factor is important in the behavior of an animal, the intensity of that environmental factor may have a profound influence. It is almost impossible to imagine experimentation in biology that did not depend upon measurement. In this respect we are completely dependent upon methods developed by the physical scientist.

STANDARDS   Any measurement is a comparison of one thing with another. If we compare *A* with *B* and then later compare *A* with *C,* the information gained can be used to compare *B* with *C*. In this simple case *A* becomes a form of standard. Modern measurement uses a set of standards which has been selected by mutual agreement. The importance of the standards in science and in commerce should be obvious.

Because we have become accustomed to certain standards, others sound strange to us. We can read that Goliath stood 6 cubits and a span and carried a spear weighing 600 shekels of iron. Was this Goliath one to inspire awe? Should his spear be feared? Not unless we have an evaluation of these standards. It turns out that Goliath was about 9 feet tall and had a 25- or 30-pound spear. Cubits and shekels could be as useful to an ancient population as our units are to us. Would the average American know any more about Goliath if he were described as about 270 cm tall?

It would be entirely possible for any laboratory to devise its own set of standards. As long as these standards were used consistently the laboratory could continue to function as an independent unit. There obviously could be very little communication between laboratories, however, and science could not have made the advances that it has. It is only

because of the universal agreement upon one set of standards that a measurement made in western United States will be the same as comparable measurements made in western Germany.

The metric system is in almost universal use in science, even though English-speaking countries continue to use another system in their commerce. Three quantities, length, mass, and time, were selected as a set of basic standards, and most other quantities are derived from these. For example, volume is length cubed, and velocity is length divided by time. Probably some other set of basic standards could be devised, but these three have worked well and have become so ingrained in our thinking that it is difficult to imagine another system.

The principal advantage of the metric system is the ease of computation. Multiplying or dividing by ten is much easier than working with a system in which a pound contains 16 oz and a typical small length is $\frac{1}{32}$ in. The system of pounds and yards was based originally on some arbitrary standards. The metric system was intended to represent certain natural features, just as the meter was a fraction of the earth's circumference. After the metric standards were established, methods of measurement improved, and it was necessary to admit that a meter was not precisely the value it was supposed to be. But the "standard meter," a unit accepted by international agreement, could still be used. Nearly all figures used in this book are expressed in metric units.

Most of the measurements in biology are essentially the same as those used in the physical sciences. Occasionally, however, biological phenomena do not lend themselves to this simple, direct expression, and it is necessary for the biologist to develop some other kind of units. An antibiotic, for example, might be a mixture of several different chemical compounds. The effectiveness of the antibiotic in preventing the growth of bacteria depends upon the source of the preparation and upon the proportion of the different compounds in the mixture. As merely referring to a certain number of milligrams of the substance does not convey adequate information, some unit which expresses the activity and can be based upon a standardized laboratory test is much more useful. Instead of an absolute amount of material we refer to an amount which will cause a certain effect. Numerous units of this sort have been used in experimentation in biology because of the variability and complexity of biological materials.

The simplest method of comparing one object with another is to place them side by side. Units of length are commonly measured in this way. A scale for weighing objects could be constructed from a simple steel spring. The heavier an object is, the more the spring will be deflected. An electrical quantity such as voltage can cause a certain deflection of the needle in a meter. The weight measured by the spring and the voltage measured by the meter are examples of determinations of the effect caused by the phenomenon being measured. Many kinds of measurements are as simple and direct as these, but in certain instances it is advantageous to make a "null" measurement. Instead of measuring a value directly we measure that amount of force required to oppose it. A balance used to measure mass consists of a pair of pans suspended on opposite ends of a beam; the material we wish to weigh exerts a force, and we add weights to counteract this force. It is easier to make precise weights than precise springs. Electrical quantities can be measured by means of a Wheatstone bridge, as shown in Chapter 13. If a null method is used, greater precision is available here also because it is easier to make and evaluate resistors than meters.

Measurements of length, mass, and time require simple units: either the basic units of the metric system or fractions or multiples of these. Certain other quantities may become exceedingly complex in that they involve two or more of the basic units and comparison with two or more standards. Force, for example, is mass times acceleration, but acceleration is change in velocity per unit time, while velocity is measured in length units per unit time. Some of the complex measurements necessary in biology are rather difficult to reduce to basic units.

One of the great difficulties in biological measurement is the response of the living material to the very act of measuring. It is theoretically impossible to measure anything without affecting it, but the physicist and chemist have learned that the effects caused by their instruments are small. Living organisms, however, respond to small changes in environment. Emotional disturbances in animals are well known, but even plants show differences in response upon handling. The experimental biologist must therefore specify the conditions under which measurements were made and be prepared to accept the fact that his figures are not normal. He can only hope that his numbers can be repeated in other measurements under similar conditions and that sets of measurements under a variety of conditions will provide him with information from which he can draw generalizations.

EXAMPLES OF MEASUREMENT

The units used in measurement may be the standard units (meters, kilograms, or seconds) or fractions or multiples thereof. Generally the fractions or multiples are chosen so that the numbers are of a convenient size. For example, 8 cm is a slightly more convenient expression for the length of my finger than 0.08 meter. Several of the units which have been given names of their own are listed in Table 4-1.

TABLE 4-1   *fractions and multiples of units*

fraction or multiple	prefix	symbol
10^6	mega-	M
10^3	kilo-	k
10^0	unit	
10^{-2}	centi-	c
10^{-3}	milli-	m
10^{-6}	micro-	μ
10^{-9}	nano-	n
10^{-12}	pico-	p

There are some exceptions in the use of the prefixes, but generally they apply to the fundamental units of length, mass, and time and to derived and electrical units, as in microwatt and kilocalorie.

LENGTH   Measurements of length are comparisons with standards, which are copies derived from the standard meter. Some measurements of length can be made more accurately now than previously. Until 1960, the standard meter was the distance between two lines scribed on a bar of platinum-iridium kept in the vault of the International Bureau of Weights and Measures in Paris. The standard meter is now defined as 1,650,763.73 times the wavelength of a specified

orange-red line in the light emitted by an isotope of krypton
(^{86}Kr). The new standard is a value believed to be constant,
and can be reproduced in any well-equipped laboratory.
By mutual agreement in 1959 among the countries using the
English system, the international yard is defined as exactly
0.9144 meter. The inch ($\frac{1}{36}$ yd) is thus exactly 25.4 mm.

A number of fractions or multiples of meters are commonly
used in measurement. These units are chosen to permit the
use of small whole numbers, rather than very large numbers
or extremely small fractions. The centimeter and millimeter
follow the prefix system described in Table 4-1, but smaller
units have names of their own. One-thousandth of a millimeter (10^{-6} m) is called a micron and is given the symbol μ.
One-thousandth of a micron (10^{-9} m) is the millimicron
(mμ), and a tenth of this (10^{-10} m) is the Angstrom unit
(A or Å). These names have a long history of use for
wavelengths of light and other small lengths. More and more
today, however, one finds "nanometer" used in place of
"millimicron."

Measurements of length are among the most familiar types
of measurements. Differences between parts of organisms,
or living cells themselves, frequently fall in a size range smaller
than is convenient for the human senses. One method of
measuring small objects is to magnify both the object and
the scale with which it is compared. Measurement with a
microscope is possible in this manner, but more commonly
we use an "ocular micrometer," a scale engraved on a
transparent disk which fits into the eyepiece of the microscope.
The optics are arranged so that the scale and the object are
seen at the same time. The ocular micrometer is calibrated;
that is, definite values are assigned to the divisions of the
scale by comparison with an accurately and finely divided
scale engraved on a microscope slide. Almost all measurements of the sizes of cells or parts of cells are accomplished
in this manner. Measurements of even smaller units are
exceedingly difficult indirect measurements and frequently
are calculated from the known geometry of an optical system.

Area and volume are derived directly from length, two or
three measurements of which enable us to calculate these
quantities. Several special units of area exist, such as the
acre and are or hectare, but only square centimeters (cm^2)
and similarly derived units are common in the experimental
laboratory. Volume can be expressed in terms of length units

cubed (cm³) if the volume is calculated geometrically from the dimensions. The liter, the volume (about 1000 cm³) occupied by 1 kg of water under certain specified conditions, is the basic metric unit of fluid volume, while the gallon (231 in.³ in the U. S.) is the commercial unit. Fractions of liters generally follow the system of Table 4-1. The microliter (μl) has sometimes been designated by the symbol lambda (λ). This practice seems to be less common now, μl being more common, but the symbol λ does exist in the literature and should be interpreted as one microliter.

MASS   Mass is a measure of the amount of material, a concept derived from Newton's second law of motion which says that force is directly proportional to acceleration. Expressed in this way, mass becomes a quantitative unit related to the qualitative idea of inertia. Weight is easily confused with mass, but weight is the force which gives a body the acceleration of gravity and thus should be expressed in units of force. Mass is a constant property of a body, while weight will vary from place to place as the acceleration of gravity varies. If we say that an object "weighs" 15 g, we really are referring to its mass.

The standard unit of mass is the international kilogram, a cylinder of platinum-iridium kept in the vault in Paris. It was originally intended to be equal to the mass of 1000 cm³ of water at 4° C, but later more precise measurements have shown that this is not exactly correct. Thus the kilogram, like the "old" meter, becomes an arbitrary standard. Since July 1, 1959, the international pound is defined as 0.45359237 kg.

Multiples larger than the kilogram are rarely used in the laboratory, and, in fact, the gram (10^{-3} kg) is probably the most commonly used unit. The prefixes and symbols of Table 4-1 apply quite directly to mass units. The microgram has sometimes been given the symbol gamma (γ), but this, like the use of λ for μl, is being discouraged.

Measurements of mass are perhaps the easiest of all measurements to make by comparing unknown masses with standard masses. We now can use a variety of balances operating on slightly different principles. The simplest idea, probably, is to place the unknown and standard masses on the opposite and equal arms of a beam. Standard weights are added until there is no deflection of the beam. The usual analytical balance is of this type. Routine weighings of larger objects might be performed on a triple beam balance, where a set of standards counteracts the unknown mass, not by changing the amount of standard mass, but by changing the distance between the standard and the point at which the beam is suspended. From the simple law of the lever the scale can be graduated in mass units. The "trip scale" is a combination of the previous two types. The material to be weighed is placed on the left pan, known weights are placed on the right pan to the nearest gram, and then the fractions of grams are found by sliding a "rider" weight along a beam.

A torsion balance contains a wire or band of metal which is twisted during the measurement. Within the range of the balance, the amount of twisting is proportional to the load. Most torsion balances are used as null instruments, being brought back to the undeflected position by adding weights to oppose the load or by moving riders. Several brands return the balance to the undeflected position by moving an arm which twists the wire in the opposite direction. The arm moves over a scale from which the unknown mass can be read.

Several new analytical balances are now being made that may eventually replace the equal-arm analytical balance. One difficulty with placing the item of unknown mass and the standard masses on opposite sides of the balance point is that the two arms of the beam may not be identical. This possibility can be avoided by comparing the unknown and standard masses on the same arm against some inert material on the opposite arm. In following such a procedure, weighing would be too cumbersome, but some of the new balances operate on this principle. A full set of standard weights is hung on one arm of the balance, counteracted by an equal mass on the opposite arm. The unknown material is placed on the same arm as the standards, and then standards are removed in various combinations until balance is restored. Recording the values of the standard weights removed gives the mass of the unknown object. These new balances fit into our mechanical world because the standard weights are removed by a system of hooks, arms, and gears, and this mechanical system is coupled to a dial which shows the mass of the unknown object directly. Those of us who grew up with the equal-arm analytical balance, however, are still a little suspicious of anything this easy.

TIME   The units in which the scientist expresses time are the everyday units: seconds, minutes, etc. The second is the standard unit, defined once as 1/86,400 of a mean solar day, more recently as 1/31,556,925.9747 of the year 1900. In 1967 a new definition of the standard second was adopted by the General Conference on Weights and Measures: "the duration of 9,192,631,770 periods of the radiation corresponding to the transition between two hyperfine levels of the fundamental state of an atom of ^{133}cesium." In other words, the new standard second is based upon an "atomic clock" which might be constructed in any laboratory.

Time is an especially important unit in biology because living material is always changing, and measurement of rates gives important information about mechanisms. Time can be measured with a clock or stopwatch, but where the human reaction time can contribute to errors some electrical or mechanical device may be used to start and stop the clock. Some exceedingly fast reactions are followed and timed by complex electronic equipment.

ENERGY   Of all the quantities that must be measured, the abstraction known as energy is probably the most difficult to understand. Energy is commonly defined as the capacity or ability to do work, but work itself is of several kinds. Energy can be converted from one form to another, and the mathematical expressions denoting these transformations can become quite complicated.

The term "heat" means many things to many people, but here we shall refer only to the quantity of thermal energy which a body contains. The amount of heat depends upon the mass of the body, and under a given set of conditions a given body of a certain material must contain the same amount of heat. Heat is rarely measured directly, but instead the amount of heat in a body is calculated from other measurements. Even the calorimeter, which comes close to measuring heat, actually measures changes in temperature of a known amount of water or other material. Heat measured in this way is expressed in calories, one calorie being the heat required to raise the temperature of one gram of water one Celsius degree. Calculating heat in joules from electrical measurements is usually more precise than calorimetry. The joule, when used for heat, is the energy given off in one second when a current of one ampere

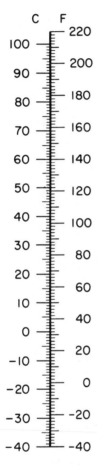

FIGURE 4-1   *Celsius (centigrade) and Fahrenheit temperature scales.*

flows through a resistance of one ohm. Thus heat can be related quite directly to electrical quantities.

Temperature is a measure of the "concentration" of heat, that is, the amount of heat per unit of material. The temperature of a uniform object is independent of its size. The Celsius * (centigrade) temperature scale is used almost exclusively in biology. Degrees on the Absolute or Kelvin scale are the same size as the centigrade degrees, but 0° K is

* The name Celsius for this temperature scale was adopted by the General Conference on Weights and Measures in 1948. The U. S. National Bureau of Standards recommends the use of Celsius rather than Centigrade (*Science* 136:254–255. 1962).

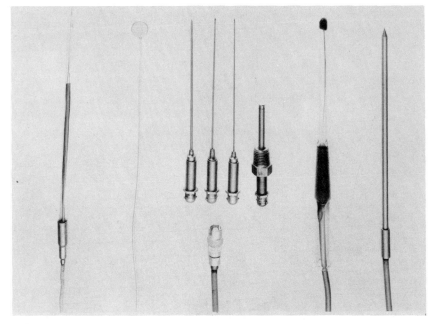

FIGURE 4-2   *Several thermistor probes, each of which can measure temperature. Size can be estimated from the wires at the bottom, each
about 3 mm in diameter. (Courtesy Yellow Springs Instrument
Co.)*

about −273° C. If temperatures must be converted from
Celsius to Fahrenheit, the easy way is to carry a diagram on
which the scales are printed side by side, as in Fig. 4-1.
Alternatively, remember a few reference points; e.g.,
20° C = 68° F, 37° C = 98.6° F, 40° C = 104° F, and
5 Celsius degrees equal 9 Fahrenheit degrees. As a last resort,

$$t_F = \frac{9}{5} t_C + 32, \quad \text{and} \quad t_C = \frac{5}{9} (t_F - 32).$$

The familiar way of measuring temperature with a thermometer depends upon the thermal expansion of mercury.
The amount of expansion per degree is constant over a wide
range of temperatures. Temperatures below the range where
mercury can be used require some other liquid, such as an
alcohol or a hydrocarbon.

Several electrical devices also measure temperature. A
thermocouple is formed by joining wires of two different
kinds. For example, a piece of copper wire could be joined

to a piece of wire of the alloy called constantan to form a loop half copper and half constantan. If one junction is at a higher temperature than the other, a small but measurable voltage will exist in the circuit, the magnitude of the voltage depending on the temperature difference.

A resistance thermometer measures temperature by detecting changes in the resistance of a platinum wire. The platinum wire can be used as one of the resistors in a Wheatstone bridge (Chapter 13), and temperatures can be measured with great precision. A somewhat similar device called a thermistor depends upon changes in resistance of a semiconducting material. Because thermistors offer the advantage that the detecting element can be extremely small, they are particularly useful in some biological experiments. Thermistors have made it possible to measure temperatures in such improbable places as stomachs. Several thermistor probes are illustrated in Fig. 4-2.

Energy exists in other forms: as electrical energy, as the kinetic energy of motion, as radiant energy, or in one of several forms of potential energy. Measurements of energy depend very strongly upon electrical instruments of one kind or another. Electrical measurements will be taken up in Chapter 13, and measurements of radiant energy are dealt with in Chapters 12 and 13. Potential energy actually cannot be measured directly but instead is calculated from the work required to transform the energy to this storage form or from the energy released when the potential energy is transformed into one of the other kinds. Examples of potential energy are the energy possessed by a body held in an unstable position and the bonding energy of chemical compounds. We might say that a molecule of glucose contains a certain amount of energy, but almost certainly we mean the energy that will be liberated if and when this molecule is converted chemically into some more stable, lower energy compound. Chemical potential energy is usually expressed in kilocalories per mole of reacting material.

CHEMICAL QUANTITIES AND CONCENTRATIONS

Chemical compounds exist as molecules, but molecules are much too small to be treated as units. A much larger unit, the mole, is Avogadro's number (6.02×10^{23}) of molecules. Avogadro's number (N) was originally devised in studies with gases when it was found that a given volume of any gas at a given temperature and pressure contains the same number of molecules. One mole of any gas at $0°$ C and a pressure of one atmosphere occupies 22.4 liters. Molecular weight is a relative figure, calculated by adding atomic weights, which in turn are determined relative to oxygen, which is given an atomic weight of 16. An amount of a compound with a mass equal to its molecular weight expressed in grams contains N molecules.

Most biological reactions occur in solutions; that is, the reacting molecules are dissolved in water or occasionally in some other solvent. Concentration is a measure of the amount of the dissolved substance (solute) in a unit volume of solution; several methods of expressing concentration are in common use. If we dissolve one gram-molecular weight of glucose (180 g) in enough water to make one liter of solution, the concentration is one mole per liter (1 M/l), which is frequently contracted to one molar (1 M). The physical chemist uses an expression, molal, for a solution in which one mole of a material is dissolved in 1000 g of solvent.

Solutions of acids which ionize to produce only one H^+ per molecule can be labeled in molar concentrations, but if the acid should liberate two or more H^+ ions, the behavior might make the acid solution seem two or more times as concentrated as it actually is. To meet this difficulty the chemist devised the notation, equivalents per liter (eq/l); a solution of 1 M H_2SO_4 contains 2 eq/l. A solution containing 1 eq/l is called a one normal (1 N) solution. These terms are also used for solutions of materials other than acids which liberate H^+ ions.

For certain purposes, concentrations expressed in percentages are adequate, particularly in the biology laboratory where molecular weights are not always known and where compounds are not always as pure as in the chemistry laboratory. Two kinds of percentage solutions are possible. A 5 per cent salt solution could be made by dissolving 5 g of NaCl in enough water to make 100 ml of solution. This is percentage by weight, and if there is likely to be any doubt, it should be so designated (w/v). A 5 per cent solution of alcohol (a liquid) contains 5 ml of alcohol made up to 100 ml with water and does not necessarily contain 5 g of solute. Such solutions *must* be designated as volume percentage solutions (v/v). Concentrations of some solutions are expressed as g/l or mg/l, and it will be seen that these are basically similar to percentage by weight. A solution expressed in milligrams per liter is sometimes spoken of as parts per million (p.p.m.), an expression which is awkward when dealing with liquid solutions but valuable when working with gases in air or other complex solutions.

The hydrogen ion (H^+) concentration, or perhaps more properly the concentration of the hydrated ion (H_3O^+), has a very profound influence on all kinds of biological reactions. Concentrations of H^+ ions might vary from more than 1 M to 10^{-10} M or less, so we use the logarithmic *pH* scale to denote these concentrations. If [H^+] is the concentration of these ions in moles per liter, then

$$pH = - \log_{10} [H^+] = \log_{10} \frac{1}{[H^+]}$$

The notation *pH* 1 means 10^{-1} M hydrogen ions, or *pH* 6.8 means the H^+ concentration is $10^{-6.8}$ M. The *pH* of a solution might be measured by noting the effect on certain dyes or, more precisely, with an electrical *pH* meter (Chapter 13).

PRESSURE   The pressure exerted by a gas is expressed in a variety of terms. The atmosphere exerts on a unit area of the earth's surface a pressure which is equal to the mass of all the air vertically over that area. Under a set of standard conditions, this pressure would be called one atmosphere. Atmospheric pressure varies greatly with temperature, however, so some more easily measurable unit of pressure is desirable. A column of mercury could be arranged so that it exerted the same pressure as the standard atmosphere. Since mercury is so much more dense than air, this column is only 760 mm high. In other words, a layer of mercury 760 mm deep weighs the same as the whole thickness of the atmosphere. If we used water, the column would be about 34 ft high. In the laboratory, pressures usually are expressed in terms of the equivalent column of mercury (mm Hg). The mass of such a column of mercury depends on its cross-sectional area. A column of mercury 1 cm² in cross section and 760 mm high weighs about one kilogram. This is roughly equivalent to 14.7 lb/in.².

Even though most laboratory work designates pressures in mm Hg, the international unit of pressure is the bar, which is equal to 10^6 dynes/cm², or 1.013 kg/cm², or 0.987 atm. Weathermen express pressure in millibars.

VOLUMETRIC Even experiments employing the most elaborate instruments
GLASSWARE are likely also to require volumetric glassware. This
glassware is a special kind of laboratory equipment, designed
for measuring the volumes of liquids. Burettes, pipettes,
volumetric flasks, and graduated cylinders are the most
commonly used pieces.

A burette is a glass tube, graduated in milliliters, with a
stopcock at the lower end so that measured volumes of liquid
can be drained off into another container. Some burettes
have special stopcocks and reservoirs to make filling easier.
When water or aqueous solutions are used in a burette, the
upper surface of the liquid forms a curve or meniscus. The
level of the liquid is measured by placing the *bottom* of this
curve at the graduation on the burette. For very precise
reading some dark material should be placed behind and
below the meniscus so that the exact bottom of the curve
will be easily identified. If the eye is placed slightly below
the graduation line, the graduation ring looks like an ellipse.
Placing the bottom of the meniscus at the center of this
ellipse gives a precise reading and avoids errors as great as
the thickness of the graduation line.

A pipette is really a miniature burette, consisting of a
glass tube with one or more graduations engraved along its
length. Measured quantities of liquid are transferred from
one container to another by sucking up a pipetteful and then
discharging the amount of liquid designated by the scale
marks. A volumetric or transfer pipette (Fig. 4-3a), designed
for transferring certain exact volumes of liquid, has only
one mark on the narrow part of the tube, where a small
difference in volume shows up as a large difference in height.
Thus these are the most precise of pipettes. The pipette is
filled to about 15 to 20 mm above the line, and then held
vertically with the index finger on top controlling the flow
of liquid. (Some beginners seem to prefer to use a thumb,
but this is about as awkward as holding a fork as you would
a hammer.) A slight rolling motion of the finger allows
perfect control, and the excess liquid is drained off down
to the mark. The liquid is then allowed to flow out at an
unrestricted rate until the level has reached the bottom of the
bulb of the pipette. Finally, the tip is held against the wet
side of the vessel until the liquid has stopped flowing. The
last portion of a drop is ordinarily not blown out of a
volumetric pipette.

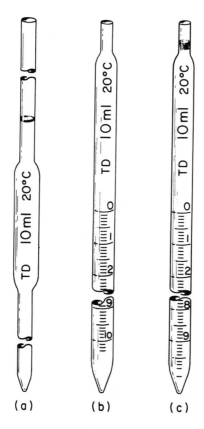

FIGURE 4-3   *Three kinds of pipettes. (a) Volumetric or transfer; (b) Measuring; (c) Serological.*

Measuring pipettes (Fig. 4-3b) are graduated in milliliters, with 0 at the top. The full capacity is contained between this mark and a mark near the bottom. The liquid is allowed to drain as rapidly as possible and still retain enough control to stop at the desired point. When the full quantity has been "delivered," touch the tip to the side of the receiving container and remove the pipette. Never blow out the last "drop" because this last "drop" may be a milliliter or more.

The serological pipette (Fig. 4-3c) is graduated to the very tip, and the last drop is to be blown out. Such pipettes are identified by a ground or etched ring at the top. Serological pipettes and measuring pipettes are superficially so similar that special care is necessary to avoid mistakes. The two types originated separately. Some biologists prefer one kind, some the other.

Biologists use pipettes more frequently than chemists do, probably for several reasons. The quantities of materials transferred are more likely to be about the right size for a pipette, and pipettes are faster than burettes. Ordinarily, more precise control is achieved with a burette, but the variation in biological materials is usually much greater than any error introduced by pipetting. Finally, the chemist is much more likely to handle acids, caustic solutions, or strong poisons, and a pipette always involves the danger of drawing up a mouthful of the liquid.

When acids, solutions of radioactive materials, or other harmful substances must be pipetted, it is essential to use a propipette rather than pipetting by mouth. These devices are of several different types, including simple rubber bulbs, rubber bulbs with valves to achieve better control, and syringes similar to hypodermic syringes.

Volumetric flasks (Fig. 4-4) are used primarily for making solutions or for diluting more concentrated solutions. A convenient method for dissolving many materials is to place the weighted quantity of solute in the flask and then to add distilled water to about $\frac{1}{2}$ or $\frac{2}{3}$ of the capacity. Shake or swirl until the solute is dissolved and then add water up to the line on the neck. Finally, pour the solution into an Erlenmeyer flask for storage. Please do not store solutions in volumetric flasks, because possible interaction between the solution and the glass may change the internal volume.

The graduated cylinder (Fig. 4-5) is the convenient piece of glassware for rapid measurements of quantities of liquid. Since the diameter is greater, relatively, than that of any other volumetric item, an error in height of liquid means a greater volume error.

CALIBRATION   This is a procedure for assigning values to the various graduations on the glassware. Ordinarily the calibration is performed by the manufacturer. Volumetric glassware is calibrated either "to contain" or "to deliver." The difference is the film of water that adheres to the glass after delivery. That is, a pipette might contain exactly 10.01 ml, but only 10.00 ml would flow out, or be delivered. Burettes and pipettes are almost always calibrated "to deliver." Volumetric flasks usually are calibrated "to contain." Graduated

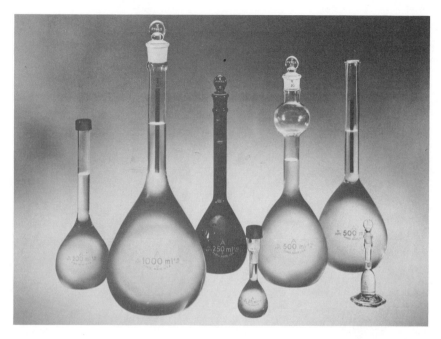

FIGURE 4-4   *Several volumetric flasks. (Courtesy Corning Glass Works.)*

cylinders might be calibrated either way. The manufacturer indicates on the glassware how it was calibrated, either with words or with the abbreviations *TC* or *TD*. Most calibrations are performed at 20° C.

Several grades of volumetric glassware are available, differing chiefly in the tolerances of calibration. Class B tolerances are usually about twice as large as Class A tolerances, as indicated in Table 4-2. As expected, Class A glassware costs more. At an even higher cost, the manufacturers provide pieces tested individually, numbered with a serial number, and accompanied by a certificate from the manufacturer's standards laboratory.

THEORY OF MEASUREMENT   Measurement, if you think of it as a process of applying numbers to an ordered sequence of units, is just counting. Some variable quantities are discrete; that is, each unit occurs individually with no fractions of units. The number of apples in a bushel or the number of people in a population illustrates such discrete numbers. The number of cents in a

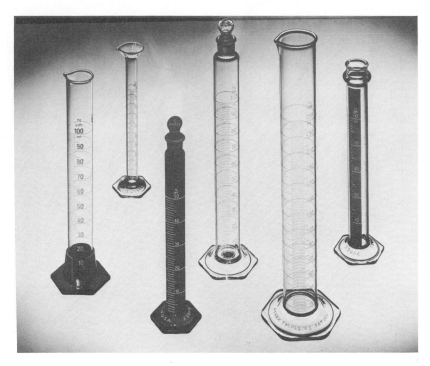

FIGURE 4-5   *Several graduated cylinders. (Courtesy Corning Glass Works.)*

certain number of dollars is also discrete, but the annual interest on one dollar at $4\frac{1}{2}$ per cent is a fractional value. The length of a room is a number of meters plus a number of centimeters plus a number of millimeters, etc. Quantities which vary in this way are called continuous. Most measurements in the laboratory deal with continuous quantities, and the counting consists of applying numbers to units of a chosen size.

We might suppose that under a given set of conditions (temperature, humidity, etc.) a bench in the laboratory possesses some actual exact value of length. If we measure the bench, we obtain an estimate of this exact value. With a meter stick we find the length to the nearest centimeter, but the true length might be a millimeter or two longer or shorter than our estimate. Using a set of optical instruments, we measure (and calculate) the length to the nearest millimeter. The estimate of the true length is better than before, but still an estimate. If better and better techniques are used,

TABLE 4-2 *tolerances of volumetric glassware: class a and class b tolerances*

capacity† ml less than and including	cylindrical graduates class a — tolerance ±ml to contain	cylindrical graduates class a — tolerance ±ml to deliver	cylindrical graduates class b — tolerance ±ml	volumetric flasks class a* — tolerance ±ml to contain	volumetric flasks class a* — tolerance ±ml to deliver	burettes class a* — tolerance ±ml	pipettes measuring class a* — tolerance ±ml	pipettes serological class a* — tolerance ±ml	pipettes transfer class a* — tolerance ±ml
0.1							0.0025	0.0025	
0.2							.004	.004	
1				0.010			.01	.01	0.006
2				.015			.01	.01	.006
3				.015					.01
4				.020					.01
5	0.04	0.05	0.08	.03	0.05	0.01	.02	.02	.01
10	.05	.06	.10	.03	.05	.02	.03	.03	.02
15									.03
20									.03
25	.11	.14	.30	.03	.05	.03	.05	.05	.03
50	.18	.22	.40	.05	.10	.05			.05
100	.24	.30	.60	.08	.15	.10			.08
200	.40	.50	1.4	.10	.20				.10
250	.40	.50	1.4	.12	.25				
500	.85	1.05	2.6	.15	.30				
1000	1.5	1.9	5.0	.30	.50				
2000	3.0	3.7	10.0	.50	1.00				
4000	5.5	6.9	18.0						

Prepared by Kontes Glass Company from National Bureau of Standards Circulars. Used with permission.
* Tolerances of Class B glassware are twice as large as Class A glassware, where not otherwise specified.
† Tolerances are established on the basis of capacity only and are independent of subdivisions.

the results approach the true value but never actually reach it.

Let us now repeat our best measurements several times. Experience teaches us that we should not expect exact duplication of results. Slight human variation, small changes in the instruments, and other fluctuations, some of which are too small to be noticed, will combine to affect the final measurement. Our series of several numbers are close to each other but not identical. If only random variations affect the results, a graph of a large series of measurements tends to follow the "normal" curve in Fig. 4-6. The higher a point on the curve, the more frequently that value is found. A sharper and steeper curve indicates a more precise measurement. Precision of measurement must be defined with reference to these random variations and refers to the closeness of agreement among the various values. In the biological laboratory, we may have to be satisfied with measurements in which the range of values, that is, the difference between the largest and smallest, is not more than 5 percent of the quantity being measured. Physical measurements are often much more precise; physicists may not be satisfied with variations as large as one part in a million.

If the measurement is accurate, the numbers obtained cluster around the real and true value. Some individual numbers will be larger, some smaller, but taken together the set of numbers gives a useful estimate of the true value. The error of a measurement is the difference between the observed value and the hypothetical true value of the quantity. If the error is small compared to the magnitude of the quantity, the measurement is said to be accurate. Measurements cannot be accurate without being reasonably precise. If, in contrast, some systematic error exists in the measuring technique, the estimate can be precise without being accurate. If you failed to notice that someone who needed a piece of maple wood had cut it from the end of your meter stick, you might make precise but inaccurate measurements.

Measurements in physics strive for precision, which is necessary for accuracy. This is the goal in biology, also, but such precision is rarely achieved because biological materials are inherently variable. Even so, the variability of biological material is no adequate reason for not making quantitative measurements. The biologist simply must realize that his measurements can never be quite as precise and, therefore, never quite as accurate as the physicist's.

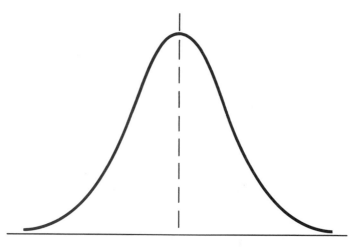

FIGURE 4-6   *The "normal curve" of error. Values being measured extend along the horizontal axis, and the height of the curve at any point is an indication of the frequency with which that value occurs.*

One might suggest that better instruments would detect smaller differences and be able to give more precise values. This is true only to a certain point. Eventually random fluctuations in atomic or molecular structure, kinetic activity of particles, or other such changes become as large as the differences we are trying to detect. The human ear, for example, is an extremely sensitive detector of slight variations in pressure. The natural kinetic or thermal movement of air molecules causes very small changes in pressure when the molecules strike the tympanic membrane. If the ear were only a trifle more sensitive it would detect the bombardment by random movement of single air molecules, and all sound would be superimposed on a steady rumble. This delicate instrument is the product of evolution, but the instruments built by man are subject to the same limitations. There is a point beyond which attempts to refine measurements are pointless.

SIGNIFICANT DIGITS   The number of significant figures resulting from a measurement is an indication of the precision of the instrument. If we weigh a pebble on a triple beam balance we find that it weighs 4.7 g. We realize that this number means that the actual mass is between 4.65 g and 4.74 g. When we use a torsion balance to weigh the same pebble, we find a weight of 4.72 g. This obviously represents a value between 4.715 and 4.724, but we cannot tell whether the pebble actually weighs slightly more or less than 4.72 g. On a good analytical balance, we might weigh to the nearest tenth of a milligram, expressing the result as 4.7208 g. These five figures are significant, revealing the precision of the measurement. It would be a mistake to weigh on the triple beam balance, obtain a weight of 7.2 g, and then write the weight as 7.2000 g. Zeroes placed after the decimal point are counted as significant figures indicative of precision. In the case of large numbers, like 563,000,000, only the first three figures have any meaning, the zeroes being required to indicate the position of the decimal point. In case of question, the number could be written as a power of ten, i.e., 5.63×10^8, and the number of significant digits shown. It is interesting to speculate on the precision required in a financial institution like a bank or the steps required to preserve accuracy and precision in taking the decennial United States Census.

More significant digits become available only by improving the measuring technique. Calculation, especially multiplication and division, tends to increase the numbers of figures, but the final result can be no more precise than the least precise of the individual measurements.

Measurements can never be more precise nor more accurate than the standards used in the measurement. I once saw a bottle of a standard acid labeled 0.100406281 N. I was told that this was prepared by weighing potassium acid phthalate to five significant figures. The solution prepared from this substance was used to standardize an alkaline solution, the alkali concentration being expressed to seven figures. The alkali was then used as a standard for the final acid solution, which became even two figures better. Possibly this acid was 0.10041 N, but certainly the extra numbers are meaningless. In fact, the weight of the potassium acid phthalate could be no better than the set of balance weights, and the several titrations can be no better than the volumetric glassware used.

Certain kinds of measurement require all the precision available. Biological materials rarely demand such precision, however, so the extra effort is wasted. It is quite possible to spend too much time in careful measurement if, for example, the living material changes during the measurement. Beginners sometimes handle solutions so slowly and with

such care that evaporation significantly changes the concentration. A reaction rate might be found by measuring
the amount of one of the reactants, say every five minutes,
but the estimate of the rate is not very good if it takes three
minutes to obtain each value. Nearly always the largest
error in any biological measurement is in the living material.
Measurements should not introduce error larger than the
biological error, but there is no point in using measurements
a good deal more precise than needed.

DIMENSIONS   Measurements of physical quantities are given in terms of
a unit, and the label defining the unit is as important as the
number. We commonly use the term dimension when referring
to these labels. Admittedly, dimension usually refers to
length, but it takes only a little imagination to think of
seconds, degrees, and grams as dimensions also.

When calculations are performed using dimensional
values, that is, numbers together with their dimensions, the
values will change but frequently the dimensions change
also. The dimensions of velocity are in terms of length units
per unit of time, or cm/sec. This relationship could also be
expressed exponentially as $cm^1 \times sec^{-1}$. If an object travels
x cm in y sec, its velocity is x/y cm/ sec.

Calculations can be shown by equations, as $A = B$,
where the "equals" sign means that A and B are identical.
If A and B have dimensions, the dimensions must be
identical as well as the accompanying numbers. Many
measurements made in the laboratory lead to expressions of
proportionality. For example, we find that a given material
obeys the relationship "mass is proportional to volume."
The relationship is true whether we know the units or not,
but we cannot write an equation M g $= V$ cm³ because
grams are not cubic centimeters. We can say, however,
M g $= kV$ cm³, if we let k be a constant. Any increase in V
brings a corresponding increase in M. Multiplication by the
constant must bring about an identity in dimensions, and
therefore k must have dimensions of its own.

$$M \text{ g} = k \text{ g/cm}^3 \times V \text{ cm}^3$$

This particular constant is called density. If M were expressed
in pounds and V in gallons, the relationship between mass
and volume would be the same, but k would have a different
numerical value and different dimensions.

Some biological measurements must be expressed in rather complex dimensions. A measurement of rate of metabolism, for example, might be in terms of microliters of oxygen used per gram of cells per hour or $\mu l \ O_2/g$ of cells $\times$ hr. It would be improper to write this as $\mu l/g/hr$ because such expressions give even the mathematician fits.

The main word of caution to be offered here is to remember the dimensions. A measurement of millimeters of pressure change is not identical to μl of O_2 unless an appropriate correction is made. If you count scale divisions on a dial, the divisions have meaning only if you know what they stand for. Biological laboratories frequently use electrical recording devices, but the record produced is in chart paper divisions until the proper dimensional corrections are applied. Calculations can be properly interpreted only if the dimensions are changed along with the numbers.

INDIRECT
MEASUREMENT

Probably even more in biology than in physics, certain measurements must be made indirectly. You cannot very well measure the concentration of a substance inside a cell by ordinary chemistry without destroying the cell. Stephen Hales measured the blood pressure of a horse directly, but his techniques cannot be used on humans.

Ingenuity on the part of the experimenter will often yield an indirect measurement of a quantity. A plant cell contains an amount of water which is not directly measurable, but generally the volume of the cell can be estimated under a microscope. It might be guessed that the amount of water is some function of the volume, $W = f(V)$, where the relationship might be proportional, logarithmic, or some more complex function. If you have reason to suppose that water content is proportional to volume, then $W = kV$, and it becomes possible to compare two cells. Even if the value of k is unknown, the results give relative values of water content.

The reasoning demonstrated is generally applicable. If you know that $A = kB$, and if you know the value of the proportionality constant, it is possible to compute the value of B from measurements of A. Even if the value of the constant is unknown, relative values for B can be obtained by measuring A.

MICROMETHODS

Biologists and biochemists have found it necessary to develop a set of microanalytical methods. These methods, in general, follow the same physical and chemical principles as ordinary analytical procedures, but working with very small quantities may introduce special difficulties. Pipetting 10 ml of water is easy, but pipetting 10 μl is more difficult because the surface tension of the water has more influence in the smaller pipette. Certain micromethods are limited by the random fluctuations or movements of molecules. Obviously special care is required when working with extremely small quantities of materials.

Some other micromethods depend upon physical and chemical principles which are not ordinarily used on a larger scale. Several of the techniques described in the later chapters qualify as micromethods by dealing with small quantities and commonly used principles or by using principles not ordinarily used on larger amounts of material.

SELECTED REFERENCES   No separate publications specifically covering measurement are recommended here.

From the Bibliography:

Wilson, E. Bright, Jr., *An Introduction to Scientific Research.* Measurements and the execution of experiments are discussed in great detail.

Richards, James A., Francis Weston Sears, M. Russell Wehr, and Mark W. Zemansky, *Modern University Physics.* Several discussions of the theory and practice of physical measurement are included.

Newman, David W., *Instrumental Methods of Experimental Biology.* J. L. Lords has a chapter on weighing devices.

Most of the modern experiments in biology employ instruments, tools, or techniques assembled from the various physical sciences. Usually several different methods or instruments are available, and the investigator must make a choice. Sometimes it is not easy to choose.

Of course the techniques to be used in a research project must be suitable and related to the problem. The most elaborate set of instruments is useless if it does not measure the right thing.

Physical science has placed many instruments at the disposal of the biologist—some simple, some extremely complex. Several features of these are noted in the following sections, and several general methods are described in detail in the chapters which follow. The method to be chosen is the one that gives the most precise and reliable information with the least difficulty and expense.

Many papers in the literature contain descriptions of complicated assemblies of parts put together for use in rather simple experiments. Such an instrument (or combination of instruments) may be unique and expensive, or it may depend on principles not commonly used. Such instruments are often constructed in the laboratories where they are used. Some people, who are born gadgeteers, do all their experimenting with such instruments. It is fun and stimulating to the imagination. Papers describing work done by such instruments are impressive, we must admit. Frequently these instruments offer the only way to make a certain kind of measurement.

Not all experimental research requires elaborate instrumentation, however. Even the most impassioned gadgeteer, with pliers for hands and vacuum tubes for brains, would agree that the most completely mechanized work is not necessarily the most "scientific." Some of the definitive solutions to biological problems have been achieved through the use of simple glassware and careful observation. Each investigator must choose for himself between the simple and the elaborate.

Another choice, not always so obvious, is the choice between a direct observation and an indirect observation. Living cells sometimes resist direct observation by the contemptible trick of dying. Even some biological chemicals can be observed only indirectly. For example, because most of the ordinary protein molecules are not visible, information about their structure must be obtained by indirect methods. It is possible to measure viscosity, density, mechanical properties, solubility in various solvents, and chemical make-up, but the picture pieced together from these bits of information will never be as complete as if the individual atoms and bonds could be seen.

SOME QUESTIONS WORTH ASKING Probably the easiest way of organizing the evaluation of an instrument or technique is to ask a series of questions about it. The following set will illustrate some features that are usually worth considering.

1. *Purpose:* For what purpose was the instrument designed? Is the technique adaptable to the present needs? Under what circumstances should this method be used? Is the instrument likely to be used for a purpose more exacting than originally intended?

2. *Theory:* What is the basic principle of the technique? What is really being measured? Which physical or chemical laws are the bases of this technique? Does it measure what we want to measure?

3. *Details:* How is the instrument constructed to accomplish its purpose? What kinds of equipment or supplies are required for this method? What does each knob control? Is a schematic diagram of the electrical or electronic system, the mechanical components, or the optical system available? Could the instrument be repaired or adjusted if necessary?

The knowledge of these details increases the pleasure in the experiment, and the operator of the instrument becomes most competent when he is familiar with them. Any unusual behavior is quickly noticed and easily diagnosed. More than one graduate student has been embarrassed during the final examination covering his thesis when one of the examiners asked what was in the "black box."

4. *Precautions:* What are the major sources of error? How can these errors be minimized? What kind of measurement should not be attempted? Does the technique give only one kind of information, or is it more versatile? Are there safety hazards?

The safety hazards deserve special mention. Because many modern instruments are electrical, severe shocks are possible. Any large capacitor might store its charge for some time after the instrument is turned off, and discharge through a human body may produce a flattening jolt. Many of the chemicals used in laboratory studies are more or less poisonous; some are extremely dangerous. Carbon monoxide is used in some metabolic experiments, and its presence in the air is not easily detected until it is too late. Poisonous materials are produced in some chemical reactions, one of which might occur as a side reaction along with the reaction being studied. Some pieces of glass apparatus are extraordinarily fragile. Hard glass such as Pyrex breaks with especially sharp edges and may give nasty lacerations. If an instrument or technique involves safety hazards, it might be well to ask "Is there a safer way to do it?"

5. *Precision:* Of what degree of precision is the technique capable? Is this precision adequate for the problem? Does this technique offer more precision than the problem requires? Would a less expensive, more convenient technique be as effective in providing answers to problems?

6. *Requirements of operation:* What must the operator know about the instrument in order to make the measurements? Does the technique require any special dexterity or any difficult manipulations? Can the operator reach all the controls easily? Does the technique contribute to excessive fatigue by causing the operator to work in uncomfortable positions? Is there a possibility of eyestrain? Is nervous tension a usual result of operating the instrument for long periods?

7. *Cost:* Will the budget stand the original expense, as well as the cost of maintenance and operating supplies? Is the instrument adaptable to various kinds of measurements? Can a higher initial cost be justified because a versatile instrument can be used for many purposes?

INSTRUMENT DESIGN   Occasionally it becomes necessary to build an instrument in the laboratory, especially if a measurement is to be made for which no commercial instrument is available. Physical scientists are more likely to build their own instruments than are biologists. The biologist does encounter this situation at times, however, and quite frequently he needs to modify a commercial instrument.

The "design" of any object, as the term is used here, refers to the complete operation of planning and drawing specifications. A properly designed instrument should give desirable answers to all the pertinent questions in the previous section. The design covers the purpose and theory; the materials to be used; the arrangement of the controls, meters, and mechanical parts; details of construction, including tolerances; and even the shape and color of the outer covering. A well-designed instrument performs the task for which it was built with accuracy, precision, and convenience. The instrument is "functionally designed" and has no "ruffles."

The design stage of instrument building might take a very long time and, if well done, often results in a superior instrument. Many home-made instruments, however, are needed immediately, so that six months or a year of designing would seem too long. Would it be better to build a "haywire" gadget that might work? Perhaps, but one must expect difficulties if theory and design are only surmised. As an example, a biologist once needed a small blower to provide a stream of air in an experimental chamber. He could find no commercial blower or pump with the proper qualifications, so he made a few rough sketches and started gathering materials to build a blower. Before actually starting construction, however, he thought it would be wise to look briefly into blower design. He found that the blower he had in mind would provide the air stream he needed, but only if it could turn almost a million revolutions per minute. In this case it was easier to redesign the whole experimental arrangement.

The question of whether to spend the effort required to build instruments is a serious one. Some biologists throw together the most outrageous assemblages of miscellaneous parts and produce equipment that performs beautifully. One biologist I know has achieved a reputation for building complicated devices that never work. Only experience and a certain natural knack enable one to decide whether to proceed to the construction stage.

The actual construction may be done by the biologist himself or by a machinist, glassblower, or some other expert. Some biologists enjoy doing their own work and gain the advantage of being able to modify the design as the work is in progress. It is desirable for the experimental biologist to know something about machine shop procedures, sheet metal work, carpentry, electronics, glassblowing, and assorted other specialties. The biologist cannot be expert in any of these fields, but at least he should know what can be done and what is impossible. Otherwise he might ask the machinist to bore a square hole, or he might ask the glassblower to repair a cracked lens without damaging the optical performance. Crude sketches assist the mechanic, but drawings prepared according to the practices of mechanical drawing, complete with all dimensions and tolerances, will make his work easier.

ASSEMBLY OF COMPONENTS   Because there is almost no limit to the types of parts and materials available today, the assembly of parts or components is limited only by the imagination of the biologist. If parts are to be made, they might be constructed of metal, wood, plastics, glass, foam rubber or plastic foam, cork, or almost any other material. Plastic or rubber tubing; assorted wires; glass tubing and ground glass joints; metal tubing with connectors and valves; and round, square, or rectangular rods of several materials are easily available commercially. Distributors stock standardized items such as screws, but also ball bearings, nylon spheres, precision gears, and an infinite variety of other specialized parts.

The parts of the apparatus may be held in place in several different ways. Scientific supply houses furnish rods and fixtures for making sturdy frames. The aluminum alloy rods are strong enough to support rather heavy components, and the connectors available allow completely flexible arrangement. Usually it is advantageous to build such a frame on a sturdy table in such a way that both front and back are accessible. Burette clamps, larger condenser clamps, and special devices for holding thermometers, heaters, and other equipment are standard parts.

Another type of supporting material consists of sheets of fiberboard with rows of holes. Parts can be attached with screws, wire, or specially made hooks. Such boards are often used to display tools, but they work as well to support laboratory apparatus. Much smaller pieces, with smaller holes more closely spaced, also are available. These small boards are useful as "breadboards" for temporary arrangements of parts.

A number of items built originally as toys have also been used in laboratories. You might arrange for a toy electric train to carry materials to inaccessible places. Children's steel construction sets provide sheets, bars, gears, and other parts that often are adequate for the purpose. Just use your imagination.

CONSTRUCTION OF
PARTS AND
PROPERTIES OF
MATERIALS

Once it has been decided to build equipment or component parts for larger assemblies, the choice of material becomes important. It is always worthwhile to study any properties of a possible material which will be important in the operations of construction and in the finished product. Many metals and alloys are available. Brass, even though it is expensive, has been found to be one of the best metals for general use in constructing small parts. It is easy to cut and machine, it can be soldered easily, and it is resistant to corrosion. If organisms are to be kept in a brass container, however, toxic levels of Cu or Zn may leach out of the metal. Aluminum is less dense, usually cheaper, but somewhat harder to work and more likely to corrode by electrolysis. One of the favorite materials is acrylic plastic (Plexiglas or Lucite), available in rods, tubes, or sheets of many sizes. It comes as the clear, transparent plastic, in a variety of colors or in opaque black. It machines beautifully on the lathe or milling machine, and pieces can be cemented together with a chlorinated hydrocarbon like chloroethane or with a cement containing the monomer from which the plastic is polymerized. For many uses this plastic is unexcelled.

GLASS APPARATUS

Glass has a number of properties which make it a desirable laboratory material. It is chemically inert, transparent, available in many forms, cheap, not too difficult to work with, etc. Glass components can be assembled with connecting pieces of flexible tubing, but for many applications ground joints are superior. The glassware manufacturers provide tapered joints or ball joints in standard, interchangeable sizes. Both are illustrated in Fig. 5-1. Table 5-1 lists some of the features of Standard Taper joints, along with some representative sizes. The joints themselves can be fused to glass tubing of almost any size. Flasks, condensers, extractors, and many other glass items are available with ground joints, so that a great variety of combinations can be assembled. Stopcocks of many designs and sizes complete the assembly. A few minutes spent examining the pictures in a glassware catalogue can be quite instructive.

Most glass apparatus made in the laboratory is formed from Pyrex tubing or rod. Handling this glass requires a hot

FIGURE 5-1   *Left, standard ball joint; right, standard taper joint. (Courtesy Corning Glass Works.)*

TABLE 5-1   *standard dimensions for full-length interchangeable taper-ground joints* *

standard joint size number ($\overline{S}$ designation)	approximate diameter at small end mm	approximate length of ground zone mm	computed diameter at large end of ground zone (gaging point) mm
7/25	5	25	7.5
10/30	7	30	10.0
12/30	9.5	30	12.5
14/35	11	35	14.5
19/38	15	38	18.8
24/40	20	40	24.0
29/42	25	42	29.2
34/45	30	45	34.5
40/50	35	50	40.0
45/50	40	50	45.0
50/50	45	50	50.0
55/50	50	50	55.0
60/50	55	50	60.0
71/60	65	60	71.0
103/60	97	60	103.0

* Prepared by Kontes Glass Company from National Bureau of Standards Circulars. Used with permission.

flame; usually a gas-oxygen torch is used. With a little practice, almost anyone can make simple items, and most laboratory biologists eventually become quite proficient. Sheets for the construction of parts of an apparatus can be cut from ordinary window glass. For very small animal or plant chambers, or for situations where optical properties are important, windows can be made from microscope slides or from the glass of 2 by 2 in. or $3\frac{1}{4}$ by 4 in. projection slides. Sheet Pyrex with good optical qualities is available and can be fused into all-glass apparatus, usually a job for a professional glassblower.

If we re-examine the information in this chapter we find a good deal of what could be called common sense. The selection of techniques or instruments, as well as the assembly of parts, becomes easier with experience. The beginner would do well to "waste" a fair amount of time thinking and making sketches before starting to build his own apparatus.

SELECTED REFERENCES   Strong, J., *Procedures in Experimental Physics*. New Jersey: Prentice-Hall, Inc., 1938. A most fascinating book, just to read. More important, it tells how to do a variety of things, from blowing glass to grinding telescope lenses.

Review of Scientific Instruments, a periodical. This journal is devoted to scientific instrumentation, and anyone contemplating a new instrument will want to check current and older issues. If you need an instrument, the chances are good that someone else has needed and built a similar instrument. You can gain from their experience.

From the Bibliography:

Wilson, E. Bright, Jr., *An Introduction to Scientific Research.* Has an excellent chapter on Design of Apparatus.

The whole concept of experimentation in biology implies that animals or plants are the subjects of these experiments. Biologists no longer experiment on "just whatever is handy." Once, when biology was young, hundreds of isolated observations, on as many kinds of organisms, all contributed to the general store of information. As biology has become a more mature science, concepts and principles have become more important. The notion of a research program suggests the existence of problems or questions that need answering. Sometimes, "whatever is handy" is exactly the right organism to provide the experimental answers to a problem. Most often, however, several kinds of plants or animals would be available. The short time spent considering the desirable and undesirable features of experimental organisms usually is time well spent.

More than once the success or failure of a program of experimental research has been determined by the choice of experimental organism. In some cases, the choice of animal or plant seems to have been a lucky or unlucky accident. In other cases, an especially fortunate choice was made by an astute, experienced scientist. Several examples of fortunate choices of organisms are given, along with one unfortunate selection that led to a notable failure, apparently entirely as a result of the whims of chance.

T. H. Morgan was one of several individuals who helped found the science of genetics. In the course of some of his first experimental attempts to learn the pattern of heredity in animals, he chose the little fruit fly, *Drosophila melanogaster,* from among more than a million members of the animal kingdom. *Drosophila* offers some immediately obvious advantages: it is small and easy to handle, it breeds rapidly (one generation about every 14 days), it produces fairly large numbers of offspring from one mating, and was soon found to grow readily under the artificial conditions of the laboratory. Later *Drosophila* was used in experiments which explained the inheritance of sex determination, a result we now know would have been much more difficult to achieve with some other organism. *Drosophila* also was used to show the relationships between chromosomes and heredity, partly because this small fly possesses a special set of "giant chromosomes." It is difficult to give enough emphasis to the contributions of *Drosophila* to advances in genetics. The collection of *Classic Papers in Genetics* demonstrates quite clearly the importance of this fortunate choice of organism.

Another organism enjoying a great reputation in experimental research is *Chlorella* (mostly *Chlorella pyrenoidosa*). Otto Warburg, the German biochemist who was one of the first great contributors in the field of cell physiology and enzyme chemistry, describes in one of his papers his need for an organism for experiments on photosynthesis. He very deliberately searched for a plant that would be easy to handle, would grow easily under artificial conditions, would carry on photosynthesis actively, and would be free of a number of annoying complications. After a number of preliminary experiments, he finally settled on this single-celled green alga, *Chlorella*. Since 1920, *Chlorella* has been used in a fantastic number of experiments, some quite remote from photosynthesis. *Chlorella* even enjoyed a period of popularity as a possible solution to the world's food problems. Even today, this simple plant is probably used in photosynthesis research, including studies of life support systems for spacecraft, almost as often as all other kinds of plants combined.

An example of an unfortunate choice of organisms also comes from research on photosynthesis. One of the first demonstrations that parts of living cells, in this case chloroplasts, can continue at least some of their activity after being separated from the rest of the cell was given by R. Hill about 1938. He ground up spinach, separated the chloroplasts by centrifugation, and then showed that when illuminated under the right conditions these isolated chloroplasts gave off oxygen. Since Hill's original experiment, this ability has been demonstrated in chloroplasts from a variety of plants and under a variety of conditions. Many years earlier, between about 1910 and 1918, Willstätter and Stoll had tried without success to detect some activity in broken-up cells of sunflower, the ordinary geranium, and a few other plants. We now know that even the best of modern technique for some reason fails to provide active chloroplasts from these species. If Willstätter and Stoll had happened to try spinach, the whole course of research in cell metabolism might have been changed drastically.

Now, what does this prove? Nothing, of course. A few isolated, spectacular examples need not be convincing, and indeed, they were chosen partly because they are exceptional.

It might be said that none of these investigators could foresee the value of a particular organism, nor the great effect it would have on future science. This is true, but *Drosophila* and *Chlorella* were chosen for a set of very good common-sense reasons. Several other available organisms had obvious disadvantages by comparison. The choice of experimental organisms is an important one, and a number of considerations are well worth weighing.

In addition to the strict scientific aspects of choosing the organism which will best answer the question, there is no harm in remembering the feelings of the general public. Experiments on animals are very necessary, but they may seem cruel to the layman who has had little experience or training and who does not understand experimental research. Sometimes a careful choice of experimental animals can make the layman feel better, without in any way damaging the experiment. Public sentiment on the care and use of experimental animals has become very strong, if we can judge from the number of bills on animal care introduced in recent sessions of Congress. Federal regulation of the care and use of laboratory animals has not yet become particularly restrictive. It is possible, however, that extremely stringent Federal regulations will be established—regulations so stringent as to hamper even medical research. This would be unfortunate because modern biologists have an appreciation of animal life that even fanatics can never condemn. They never condone needless cruelty. For personal reasons, as well as for the reason that it makes experimental results more reliable, the biologist treats his animals extremely well.

The following paragraphs describe some of the criteria which must be remembered in choosing experimental organisms. This list is divided into two groups: first, those features that are absolutely essential, and, second, those features that are desirable but do not make the experiment impossible if these features are not present. Most of the essential features are probably quite obvious, but even these occasionally are forgotten. Forgetting to consider these features in advance might lead one to a rather embarrassing situation.

**ESSENTIAL
FEATURES OF
EXPERIMENTAL
ORGANISMS:**
COMPATIBILITY
WITH THE
PROBLEM

If some special activity of living organisms is to be studied, then it is important to be sure that this special activity occurs in the organism. Often the problem or hypothesis to be investigated was suggested by previous work with some organism, and it is only natural to proceed with the same animal or plant. Especially when one is beginning in a new area, however, or with a new kind of organism, it is well to perform some preliminary experiments to make sure that the process one is interested in actually occurs in the organism chosen.

COMPATIBILITY
WITH THE
TECHNIQUES OF
INVESTIGATION

Most experimental research employs chemical methods or physical instruments. Is it possible to use the available methods or instruments upon the organism? A negative answer to this question might be obviated in either of two ways: choose a different technique, or select a different organism. Sometimes an animal or plant is too large or too small. Often special structural features of the cells or higher organizational units make a particular organism impractical. If one kind of plant has cell walls so thick that it is difficult to release the cell contents, perhaps another species should be sought. If the cells of one animal contain so much of one amino acid that it interferes with the analysis of other amino acids, try another animal.

AVAILABILITY

As a rule, organisms to be used in experiments should be easily obtainable. This may involve collecting the animals or plants in the vicinity of the laboratory. If the species being investigated is a rare one, the research may be seriously hindered. Difficulties along this line are being experienced in some of the marine laboratories. The worm *Urechis,* a longtime favorite at west coast laboratories, is difficult to find where it used to be plentiful. One means of avoiding these difficulties is to select an animal or plant that is a common item in commerce. If the pronghorn antelope and the sheep both show a feature which is to be studied, then use the sheep. If either wheat or buffalo grass would be suitable, choose the wheat because seeds are more readily available. The cost of the living material is to be considered also. In some studies it is desirable to raise the experimental organisms within the research facilities. Many laboratories maintain colonies of mice, rats, or rabbits because they can produce more uniform animals at any desired time. Many plant investigations are carried out with plants that have been raised in a greenhouse, or better, in chambers where all growth conditions can be carefully controlled. To summarize, then, an organism for experimental purposes must be easily collected in the vicinity of the lab, must be cheap, or must be susceptible to cultivation or maintenance under artificial conditions. Lest it be thought that this consideration is very obvious, let it be known that the present author was once involved in a study using marine seaweeds in a laboratory in Minnesota, more than a thousand miles from the nearest ocean.

DURABILITY   Another less obvious essential feature of the experimental organism is durability. Some animals or plants are better able to stand the experimental treatments than others. If a certain organism is extremely susceptible to temperature changes, and it is impossible to maintain a constant temperature, perhaps another organism might serve as well. If a long-term experiment is planned, the natural life span of the organism should be considered. Organisms occasionally confound an experiment by dying anyway, and one would be foolish to increase the likelihood of this disaster by choosing a fragile species.

SPECIES   A feature that is less directly related to the experiment, but
IDENTIFICATION   is of paramount importance in the subsequent commu- nications, is the name of the species. Different species of organisms, or even members of the same species from different sources, may differ markedly in their responses to experimental treatment. If others are to be able to repeat the experiments, it must be possible for them to identify the organisms.

DESIRABLE   If either a well-known species or a rare species can be used,
FEATURES: the well-known species should be chosen. In this respect,
GENERAL   there may be some economic or political advantage in using
KNOWLEDGE   the well-known species. One of the reasons for using sugar
OF THE SPECIES   beets in our own laboratories is that the sugar beet is an important crop in our area. It is somewhat easier to justify the expenditure of funds on sugar beets than on dandelions.

There is no benefit to be gained from the use of an exotic species *because* it is exotic.

BACKGROUND
INFORMATION

Some species of animals and plants have been used for many studies in the past, and a great volume of information has been collected. This background information may offer an advantage over a less widely studied form. The laboratory rat and mouse, the rabbit, corn (maize), *Chlorella,* the bacterium *Escherichia coli,* and several other species are all covered by a tremendous amount of literature. If there is any question about the compatibility of a certain organism with the problem under investigation, a search through this literature may provide the answer.

GENETIC
BACKGROUND

The genetic background of the animals or plants is likely to be extremely important. Usually genetic homogeneity is desirable in that it decreases the natural variation among the organisms. Because the rats and mice in common labo-ratory use have been inbred for many generations, there is relatively little genetic variation. Organisms with little genetic variability are not as strong as "wild types," however, and, in at least one of the strains of laboratory rats, "wild rats" are introduced at planned intervals to maintain vigorous stock.

If records of genetic background or pedigrees are known, so much the better. One of the largest groups of pedigreed dogs is maintained at the University of Utah for use in studies on the effects of radiation. Since each dog's ancestors are known for several generations, it is possible to relate inheritance with susceptibility to radiation damage.

AVOIDANCE OF
UNNECESSARY
COMPLICATIONS

Usually biological experiments are difficult enough without the addition of unnecessary complications. If it is possible at all, it is desirable to avoid these complications. Many of the early investigations on cell metabolism used yeast cells because they are simple, complete organisms. One of the advantages of *Chlorella* in photosynthesis studies is that each cell is a unit in contact with the experimental environment. There is almost no direct interaction between cells. In higher plants, and even more in higher animals, the internal orga-nization is so complex and there is such close interrelationship between cells that the interpretation of experimental results is often difficult.

REPRESENTATION
OF GENERAL
GROUP

It is desirable that the experimental organism be represen-tative of a general group. Rats are more-or-less typical mammals; sunflowers are typical flowering plants. Exceptional behavior may be interesting, but it may contribute little to our understanding of general biological principles. If we are to induce the general principles from the experimental results, the ordinary is preferable to the unique. As an example, a great to-do is raised about insectivorous plants although they are relatively rare (except in pictures in biology texts). Experiments have been performed on these plants, but they contribute only to our knowledge of insec-tivorous plants and very little to our understanding of general botanical principles. A process that can be studied in sunflower or soybeans, however, leads to much broader generalizations. Even in pure, fundamental, basic research, where problems are investigated only to increase knowledge, most of the problems are "human problems." Any study of an unusual or unique organism may be justified as satisfying curiosity, but the results often stand alone as details and may never be incorporated into the general advance of science.

PREPARATION
OF ORGANISMS
FOR EXPERIMENT

This section includes a number of ideas important to the preparation of biological materials for experiments. The list is intended as a set of examples and certainly cannot describe in detail the methods that are in use. Often the investigator's ingenuity is required to prepare his own organism for his own experiment.

One very important consideration in the handling of experimental organisms is the treatment *before* the experi-ment. Failure to consider the previous treatment can lead to disastrous results. A favorite examination question of some professors describes two sets of organisms that show quite different behavior in apparently identical circumstances. The desired answer or explanation is that the previous history has influenced the responses. In many laboratories, plants are raised in special insulated chambers where temperature is controlled, humidity may be controlled, illumination is provided by a color-balanced set of fluorescent tubes, lights are turned off and on by a clock, and the plants are watered with a mineral solution of known composition. Plants raised under these conditions show much less natural variation than those raised in a greenhouse or out of doors.

Animals deserve and require elaborate handling. Certain

minimum standards must be met in providing comfortable quarters, adequate sanitation, food and water, and cages that are large enough in order to maintain the animals in good health. Certainly, you would expect that one of the assumptions in animal experiments is that the animals are in good health. Diseased animals necessarily produce abnormal results. Several useful manuals and pamphlets on the general care of animals are included in the references at the end of this chapter.

In the selection of organisms for experiment, it is commonly possible to consider the use of only part of an organism rather than the whole organism. The same general rules that govern the selection of species apply to the selection of parts of organisms, although there is now the added assumption that the removal of the part for study has not altered the activity being studied. It is convenient, however, if not philosophically sound, to assume that the separation of parts makes relatively little difference in the processes being measured. Common sense, good judgment, and experience can be used to advantage in deciding to use a part of an organism instead of the whole organism.

PREPARATION OF
PLANT PARTS

The preparation of plant parts often is easier than comparable preparations of animal parts. Many of the processes of plants proceed in each cell, and the degree of coordination is lower than in animals. In dissecting out a muscle we must worry about the coordination with the circulatory system and the nervous system. If a leaf is removed from a plant, however, there may be relatively little effect on either the plant or the leaf. A number of investigations have shown some differences, usually stimulations, in the respiration of isolated plant parts, but these are differences in degree rather than in quality. Young leaves and old leaves are never identical, but it is not difficult to select leaves of about the same morphological age.

Leaves or pieces of leaves are well adapted to measurements of metabolic activities. If we wish to express the rate of oxygen production per milligram of green tissue, we merely weigh the pieces of leaf tissue. The shape need not concern us. The measurements sometimes are more meaningful when related to leaf area. If we must know the area, some shapes are more desirable than others. A cork borer easily punches out uniform discs of leaf tissue. By carefully selecting a cork borer of 1.12 cm diameter, we get discs with an area of 1 cm². Determination of the total area of leaf tissue thus requires no calculation, only counting.

Sections of other tissues can be cut in the same way. Storage roots or tubers are often quite uniform in cellular composition and have been used to advantage in many experiments. A cork borer conveniently punches out cylindrical plugs of such tissues. The plugs may then be sliced to any desired thickness so that the volume of tissue is calculated easily. Alternatively the cork borer will cut through a stack of tissue slices to yield the same result. No extra effort is required, and the number of calculations is reduced. This, in turn, decreases the chance of computational errors.

If the process under investigation demands it, whole leaves, stems, roots, flowers, or fruits may be used. Stems, leaves, or roots of germinating seeds are useful because at this stage the plant is more actively performing a variety of functions than at any other time. Often there is a choice of several kinds of plants. If one species will provide the desired organs more conveniently than another species without any attendant disadvantage, then by all means select the convenient species.

PREPARATION OF ANIMAL TISSUES Usually more care must be used in the preparation of animal tissues because of the more highly coordinated organization of the animal. Changes in the acidity, in the concentration of oxygen or other gases, or in the osmotic balance in the vicinity of a tissue may alter its behavior markedly. With proper attention to such details, however, many different animal tissues will continue to function for some time after being removed from the animal.

Before any such dissections are attempted, it is imperative to immobilize the animal somehow. Rats, rabbits, and other mammals can be anesthetized with barbiturates or similar drugs. Frogs may be immobilized by destroying the brain and spinal cord, or by anesthesis with urethane. The manner in which animals are handled is important because they can become emotionally excited.

Of all the various animal tissues, perhaps muscle has been used as commonly as any other for physiological experiments. Pigeon breast muscle was exceedingly important in unraveling the sequence of reactions in cellular respiration. Much of what we know of movement and reflex action was learned from experiments on isolated muscles. Liver has also been used very extensively, partly because this easily accessible organ provides a large mass of fairly uniform tissue, and partly because this organ is extremely active metabolically. Nerve tissue, such as rat brain, may be desirable because the metabolic rates are very high. Under proper conditions, almost any animal tissue will continue to perform some functions for a while after removal from the animal. Since osmotic relationships are extremely important, the tissues are often bathed in a solution of various salts having approximately the same composition as the liquid which normally bathes them.

MICROORGANISMS   An experiment can be made more convenient by eliminating other activities that proceed simultaneously, a purpose achieved by the use of a part of an animal or plant. The limit in simplification is reached when single-celled organisms, or microorganisms, are used. Here there can be no interaction between cells as in the higher organisms.

Algae, fungi, bacteria, and protozoa of various kinds may be very desirable experimental organisms. The contribution of *Chlorella* to photosynthesis was mentioned earlier. The single green cells grow rapidly in artificial nutrient solutions. Among the fungi, ordinary baker's yeast is undoubtedly the most convenient. Any grocery store sells yeast cakes which are pure cultures of *Saccharomyces cereviseae*. In any of the larger cities, the same yeast is available by the pound. If the yeast cake contains more starch and other substances than is desirable, the cells are washed by suspending them in water or a buffer solution and then centrifuging at low speed. The cells will settle, leaving most of the starch and the soluble materials in the liquid.

Amoeba, Paramecium, and several other single-celled animals have been used. One which has become increasingly popular in the past several years is *Tetrahymena pyriformis.* It is a tough little animal that grows well under artificial conditions and offers some decided advantages in experiments on metabolism.

Microorganisms are raised in the laboratory in "pure culture"; that is, the culture contains the one desired organism and no other species. A pure culture of *Chlorella* contains the minute green spheres characteristic of this species and presumably no bacteria, no protozoa, and no other species of algae. The original separation of a pure culture is a tedious task because microorganisms almost never exist singly in nature. Once a pure culture is achieved, it must be maintained in such a way that it will not become contaminated with the foreign organisms which are present everywhere.

The techniques by which microorganisms are raised and maintained are so complex that they can only be learned first hand. Every biologist should have at least one good laboratory course in bacteriology or microbiology. Only a brief, very general description can be provided here. Every organism must be supplied with certain chemical materials in order to grow and multiply. Microorganisms commonly are grown on artificially prepared media which will provide a source of carbon, water, nitrogen, several mineral salts, oxygen or other gases, and vitamins or any other required compounds. Sometimes this medium is a liquid; other times it is semi-solid because of the addition of gelatin or agar. If the microorganism is autotrophic and can make its own carbon compounds from carbon dioxide, it must be provided with light or another energy source. The sterile handling and aseptic techniques require constant practice. More than one experiment has been ruined or delayed by the presence of foreign organisms in a supposedly pure culture. This type of disaster becomes even more dangerous if the investigator is not aware of the invasion by the other species. Usually a routine microscopic examination and other checks are worth the few minutes required.

Occasionally experiments are performed with "enrichment cultures" of microorganisms, or, as they are sometimes called, "almost pure cultures." Any such experiment is subject to some question—especially if the proportion of the various species is changed by the growth during several transfers to new media—because the measured responses might be produced by the other organisms present.

I cannot urge strongly enough that every biologist learn the sterile techniques required in bacteriology. Of course they apply directly to microorganisms, but the general principles apply to the handling of the large plants and animals as well.

TISSUE CULTURE Some kinds of cells or tissues can, if properly handled, continue to grow after being removed from the higher plant or animal. A solution or other growth medium is prepared which will contain all the chemical compounds the cells normally require. Most of these solutions are very complex mixtures, often containing extracts of poorly known composition from living or dead cells. A representative solution for the growth of isolated plant roots would supply sugar, mineral salts, and vitamins which normally are made in the leaves, and perhaps other compounds.

The cultivation of animal tissues requires carefully prepared media, attention to supplies of oxygen, and controlled temperature. Most cells in higher animals have become so highly specialized that they have lost the power to divide and therefore are unsuitable for tissue culture. Embryonic cells or cells from malignant tumors are the most commonly used animal materials. Certain advantages are offered by animal cells raised in an artificial medium. The cells usually do not become specialized, but remain "young" and metabolically active. Tissue culture cells can provide a convenient medium in which to raise viruses, which can be cultivated only in living cells. The exact causes for the differentiation and specialization of cells in animals are not understood. Tissue culture cells, therefore, are a potent field for investigation in their own right.

Plants normally grow at the tips of roots and stems as long as they are alive. These growing tips can be cut off and will continue to grow in the proper medium. Stem tips will produce new shoots and roots and become new plants, but roots continue to grow as roots almost indefinitely. P. R. White maintained a culture of tomato roots for several years, using them in a variety of experiments.

Plant or animal tissue culture is not easy and requires the same aseptic handling as the culture of microorganisms. Mutation is always possible in either microorganisms or tissue culture. Occasionally an experiment calls for special living materials, however, and tissue culture may be worth the effort.

PREPARATION OF PARTS OF CELLS   In present-day research, many of the important problems are those involving the processes associated with all cells. More and more experiments have been performed on isolated parts of cells. Until relatively recently, it was believed that cells died immediately, completely, and irreversibly upon being broken. It is now possible to isolate almost any of the parts of cells and keep them functional at least for a short time. Nuclei have been removed from *Amoeba* cells, leaving only the cytoplasm. The cytoplasm alone cannot grow or divide, but many of the vital activities continue. Nuclei and even chromosomes have been separated in fairly large quantities from a variety of plant and animal tissues. Chloroplasts separated from green plant cells carry on an amazingly complex chemistry. Mitochondria have been isolated and continue to metabolize. Even the ribosomes, which are below the limits of visibility of the light microscope, continue to perform chemical reactions if properly treated. Several hundred enzymes have been purified from biological materials. Many of these purified enzymes can be handled like any other chemical materials.

A set of routine steps has evolved for the separation of almost any of these subcellular particulates (particles of cellular material below the cellular level of organization). The procedure used for the preparation of one particle obviously will not work for the preparation of another. Even the steps involved in the preparation of one entity, such as mitochondria, are different from one laboratory to another. As you might expect, particulates prepared by one method do not show exactly the same kinds of behavior as particulates prepared by another method. If experiments are to be repeated, the preparation of the parts of cells must follow the same routine as nearly as possible.

Nevertheless, all of the various procedures involve many of the same steps, including grinding and differential centrifugation. The differences between methods are differences in composition of solutions, in the time and speed of centrifugation, and in the order in which operations are performed. It seems reasonable to present a general discussion of the various steps, followed by one specific technique as an example. References at the end of the chapter should make it possible to find techniques for preparing other subcellular particulates.

ASSAY METHOD   In most isolation procedures it is necessary to make sure that the particles retain their chemical abilities by making measurements of activity at various stages of the preparation. If it is suspected that the particles will catalyze a certain chemical reaction, the assay consists of measuring the rate of that reaction. After a routine procedure has been developed, the assay can be carried out at less frequent intervals or delayed until the end of the preparative steps.

TEMPERATURE   Most of the preparative steps are conducted at low temperatures. The parts of cells are unstable anyway, breaking down very rapidly at room temperature. The most convenient means of maintaining a temperature of 0 to 4° C is to work in a walk-in refrigerator. Most laboratories conducting cellular physiology and biochemistry experiments now have at least one such cold room.

GRINDING   The tissue is usually broken up first by some mechanical method. A mortar and pestle offer a gentle means of grinding. The tissue is ground together with some water, buffer solution, or a complex solution of sugars, salts, etc. Sometimes sharp sand is added to speed the grinding. Another grinding device, almost as gentle, is the homogenizer, constructed of a hollow glass cylinder with a closely fitted cylindrical pestle. Several models are on the market. Larger quantities of material are chopped finely in a blender such as those used in kitchens. The high-speed blades can macerate almost any plant or animal material (including fingers) in a short time.

High-frequency sound waves, often called "supersonic" or "ultrasonic" vibrations, can be applied to some cells with reasonable success. An oscillator serves as a source of sound waves which are transmitted through oil to the container in which the biological material is placed.

Some cells have particularly tough cell walls which are difficult to break without destroying the cytoplasmic particulates. In this case, it may be feasible to use a suitable enzyme to digest away the cell wall. Lysozyme is frequently used for this purpose with cells of *Escherichia coli* and other microorganisms.

SEPARATION OF PARTICULATES   Most of the grinding procedures result in a thin paste or fluid suspension containing a mixture of cell parts and other materials. Most such suspensions contain unbroken cells, cell walls, pieces of connective tissue, and other odds and ends. Straining through a coarse filter material removes much of this unwanted matter. Filter materials commonly used are muslin, pads of glass wool, cheesecloth, and facial tissue supported by cheesecloth.

Further separation of the particulates depends upon centrifugation. Spinning at a low speed will cause settling of particles more dense than those desired, leaving the desired particles in suspension. The liquid is poured off and centrifuged at a higher speed. During this centrifugation the desired particles settle out, leaving any less dense particles suspended in the liquid. The sediment is resuspended in a clean solution, ready for use. Or it can be "washed" by recentrifuging and then is finally resuspended in the desired solution.

PRESERVATION OF ISOLATED PARTICULATES

The procedures involved in the preparation of subcellular particles are usually quite time consuming. It is often impractical to prepare a new batch of particles for each experiment, and yet, under ordinary conditions, the particles may not survive from one day to the next. Fortunately, as many of these materials can be preserved by freezing, large amounts may be prepared at one time. Chloroplasts frozen and stored at $-40°$ C have maintained their activity until the supply was exhausted, a period of many months.

The freeze-drying technique, sometimes called "lyophilization," freezes and dries at the same time. The material to be treated is placed in a container that can be evacuated. One wall of the chamber is in contact with solid carbon dioxide at about $-60°$ C. The reduced pressure causes the evaporation of water from the biological material at the expense of heat from the material. This water is then trapped by freezing against the cold wall of the container. The resulting dry material can be stored for long periods and then reconstituted by adding water. Some materials preserved in this way retain most of their original activity.

FLOW CHARTS

A flow chart such as that shown in Figure 6-1 is a useful device for describing the separation steps. In addition to helping a reader to understand the procedure, the flow chart has the added benefit of systematizing one's own thinking about the separation.

PREPARATION OF CHLOROPLAST FRAGMENTS —AN EXAMPLE

The following example is only one of many methods of preparing chloroplasts or chloroplast fragments. These chloroplasts were once prepared by Dr. J. D. Spikes and his colleagues at the University of Utah for the principal purpose of studying the light-absorbing phase of photosynthesis. Much of the other activity normally carried on in chloroplasts does not occur in these fragments. Although this inactivity would be unfortunate in some circumstances, it is

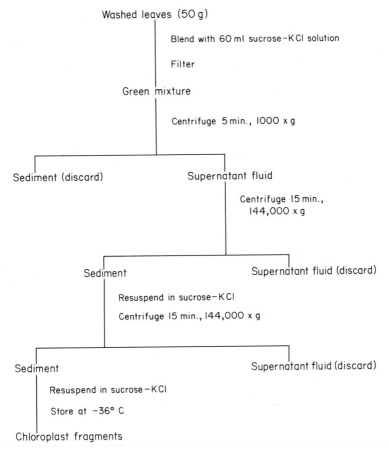

Washed leaves (50 g)

Blend with 60 ml sucrose–KCl solution

Filter

Green mixture

Centrifuge 5 min., 1000 x g

Sediment (discard)   Supernatant fluid

Centrifuge 15 min.,
144,000 x g

Sediment   Supernatant fluid (discard)

Resuspend in sucrose–KCl

Centrifuge 15 min., 144,000 x g

Sediment   Supernatant fluid (discard)

Resuspend in sucrose–KCl

Store at −36° C

Chloroplast fragments

FIGURE 6-1   *A sample flow chart for the preparation of chloroplast fragments.*

advantageous here because it simplifies the experimental setup.

Other chloroplast preparations can perform other activities, but the preparative steps, of course, are slightly different. If the chloroplasts are prepared according to one set of prescribed steps, they can carry on phosphorylation reactions in the light and may even fix some carbon dioxide.

In Dr. Spikes' method, sugar beet leaves are the usual source of chloroplasts. The sugar beet plants are raised in a special chamber under controlled conditions of light and temperature. The leaves are harvested and washed in cold water. In the cold room, approximately 50 g of washed leaves are placed in a Waring Blendor with 60 ml of a solution

0.5 M in sucrose and 0.01 M in KCl. The mixture is blended for two minutes, resulting in a thin paste which is filtered through four layers of cheesecloth. The green liquid is centrifuged for five minutes at 1000 × g. The supernatant fluid, consisting of a mixture of chloroplast fragments and other cytoplasmic materials, is separated from the sediment of whole cells, nuclei, and cell walls. The fluid is then centrifuged in a preparative ultracentrifuge for fifteen minutes at 144,000 × g. The supernatant fluid is discarded, and the sedimented chloroplast fragments are resuspended in the sucrose-KCl solution and centrifuged again. Several such washing steps are carried out, each of which removes much of any remaining cytoplasmic material other than chloroplast fragments. After the final centrifugation the chloroplast fragments are suspended in sucrose-KCl solution.

Chlorophyll in the chloroplast fragments is determined spectrophotometrically by comparison with a standard curve. The suspensions are diluted with sucrose-KCl solution until they have approximately 500 mg of chlorophyll per liter. The final dark green suspension is stored in small test tubes in the freezer at −36° C. At this storage temperature the chloroplast fragments retain their photochemical activity for months. The small tubes are removed from the freezer and thawed for use in experiments as needed.

This, then, is one method. Others will be found in the following references.

PROBLEM   1. A tremendous variety of small organisms can be found in the water of ponds and streams. Many of these organisms will grow under laboratory conditions. Select an organism that appeals to you and prepare a pure culture of this species.

SELECTED REFERENCES   Arnon, D. I., "Cell-free photosynthesis and the energy conversion process," in *A Symposium on Light and Life,* William D. McElroy and Bentley Glass, eds. Baltimore: Johns Hopkins Press, 1961. pp. 489–569. A review which discusses (among other things) the problems in the use of isolated chloroplasts. Provides references to most of the methods of preparing chloroplasts.

Brachet, Jean, and Alfred E. Mirsky, eds., *The Cell: Biochemistry, Physiology, Morphology,* vols. I–V. New York:

Academic Press, 1959–61. A comprehensive, authoritative reference work: vol. I (1959)—Methods and problems in biology; vol. II (1961)—Cells and their component parts; vol. III (1961)—Meiosis and mitosis; vols. IV and V (1960–61)—Specialized cells.

Cass, Jules S., Irene R. Campbell, and Lilli Lange, *A Guide to Production, Care and Use of Laboratory Animals: An Annotated Bibliography.* Federation Proceedings 19(4) Part III, Supplement No. 6. 1960. A list of references to the older literature with abstracts of the papers. Information on normal animals, disease, nutrition, breeding programs, and most of the other topics that might concern the experimentalist.

Crowcroft, Peter, *Mice All Over.* Chester Springs, Pa.: Dufour, 1966. A delightful story of life with too many mice.

Gray, William I., ed., *Methods of Animal Experimentation,* volume I. New York: Academic Press, 1965. Fundamental and well-developed techniques.

Harrigan, W. F. and M. E. McCance, *Laboratory Methods in Microbiology.* New York: Academic Press, 1966.

Hayashi, Teru, ed., *Subcellular Particles.* New York: The Ronald Press Company, 1959. Twenty contributors discuss the relationships of structure and function, as demonstrated by modern techniques.

Mercer, Frank, 1960. The submicroscopic structure of the cell. *Ann. Rev. of Plant Physiol.* 11:1–24. An especially pertinent review on the preparations, characterization, experimental use, and interpretations of parts of cells.

U. F. A. W. staff, eds., *The U. F. A. W. Handbook on the Care and Management of Laboratory Animals,* 3rd ed. London: Universities Federation for Animal Welfare, 1967. Published by this British society; the single most complete handbook on the subject.

Umbreit, W. W., R. H. Burris, and J. F. Stauffer, *Manometric Techniques,* 4th ed. Minneapolis: Burgess Publishing Co., 1964. Includes an excellent chapter on the preparation of subcellular particulates.

Willmer, E. N., ed., *Cells and Tissues in Culture.* New York: Academic Press, 1965. A three-volume set.

A centrifuge is an instrument designed to separate materials of different density from each other by virtue of a centrifugal force. Since the centrifugal force is similar in its effects to gravity, most things that can be separated in a centrifuge would eventually settle because of gravity, but a very long time might be required. The centrifuge allows us to hasten this effect by applying a larger force.

In laboratory centrifugation, at least one of the components to be separated is a liquid. The other might be solid particles, another liquid, or, rarely, even bubbles of gas. The separation of parts of cells, as described in the previous chapter, is one common use of the centrifuge. So many other kinds of mixtures must be separated in the laboratory routine, that a centrifuge is used almost daily. In addition to its use in preparing materials, the centrifuge is a valuable analytical tool.

The usual centrifuge consists of a rotor or head, driven by a motor. Several types of rotors are available. Some hold only a few very small tubes or vials of material, while others hold bottles with a total capacity of a liter or more. As the rotor turns, the liquid and its suspended material are subjected to the centrifugal force. The various rotors in ordinary use fall into two general classes: those in which tubes of liquid are held firmly at some fixed angle (like 35°), and those in which tubes or bottles are placed in metal buckets which swing out to a horizontal position as the rotor turns. Each type has certain advantages.

CENTRIFUGAL
FORCE

Any rotating body is subject to a constant acceleration inward, toward the center of the circle. A weight whirled on a string must be pulled inward constantly to prevent the weight from taking its natural course, that is, flying off at a tangent. This inward force, which accelerates the mass toward the center of the circle, is centripetal force. Equal and opposite to it is the outward centrifugal force. The centripetal force happens to be easier to calculate.

The magnitude of the force depends upon the speed of rotation and upon the radius of the circle. If a wheel is turning with an angular velocity of ω radians per second (a radian is the portion of the circumference of a circle equal in length to the radius R), the velocity (v) of a point on the surface is $v = \omega R$. The velocity (in units of length per unit of time) does not change, but because the point on the surface of the circle is constantly changing direction, the point is subjected to an acceleration $\alpha = \omega^2 R$. The centripetal force F_c is the mass (m) times the acceleration, or $F_c = m \times \omega^2 R$. The angular velocity ($\omega$) can be converted to revolutions per second because 2π radians is one full circle. The centripetal force then becomes

$$F_c = m(2\pi N)^2 R = m 4\pi^2 N^2 R$$

where N is revolutions per second. Centrifugal force is equal in magnitude.

The amount of force relative to gravity is a more useful figure than this absolute F_c. Usually a relative centrifugal force (RCF) is calculated by dividing by the force of gravity. The relative centrifugal force is expressed as "so many times g" or "so many g's." The force of gravity is mass times the acceleration of gravity (980 cm/sec^2) so

$$\mathrm{RCF} = \frac{F_c}{F_g} = \frac{m 4\pi^2 N^2 R}{m \ \mathrm{g}}$$

If we know R and can measure N, we can calculate the g's. Notice that the mass, m, cancels out of the equation. The values 4, π^2, and R (for any given rotor) are all constant. Therefore, the RCF is proportional to the square of the speed of rotation. Any new centrifuge must be calibrated if we are to describe its performance adequately. We measure N with a tachometer or stroboscope. These instruments usually give revolutions per minute, so $N = \mathrm{rpm}/60$. Figure 7-1 shows the results of the calibration of a centrifuge.

ANGLE HEADS   The heads or rotors in which the tubes are held at a fixed angle develop a higher *apparent* centrifugal force than the swinging bucket rotors. As shown in Fig. 7-2, particles moving downward (away from the center) must move against the viscosity of the liquid in which they are suspended. If the tube is inclined, the distance the particles must move against this counterforce is only the distance across the tube instead of the full length. The particles in the angle head move across the tube and then slide down the wall against less resistance. The effect on the settling of particles is the same as increasing the relative centrifugal force. The initial settling is faster and more thorough. The final result of a

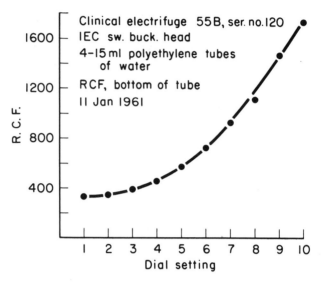

Clinical electrifuge 55B, ser. no.120
IEC sw. buck. head
4-15 ml polyethylene tubes
    of water
RCF, bottom of tube
11 Jan 1961

FIGURE 7-1   *Typical calibration curve for a laboratory centrifuge; student data.*

long centrifugation, of course, must be the same as in the horizontal tube. The chief advantage is the reduction in time required for adequate separation. The *apparent* RCF can be calculated from a complex equation, but usually it is easier to take the manufacturer's word for it.

TYPES OF   Such a variety of centrifuges is available that it is impossible
CENTRIFUGES   to describe them all. Probably the easiest way to gain information about the types and sizes is to read catalogue descriptions and advertisements. Even any attempt to classify the various types is completely arbitrary, but for the sake of discussion they can be divided into clinical centrifuges, "high-speed" centrifuges, and preparative and analytical ultracentrifuges. In common usage any centrifuge which turns faster than about 20,000 rpm is now called an ultracentrifuge.

The clinical centrifuge is a small, portable model, easily used on a bench-top. The capacity is usually not more than about 200 ml, and relative centrifugal forces up to about 2000 $\times$ g can be attained. A variety of swinging bucket and conical or angle heads can be used. Also, several special rotors

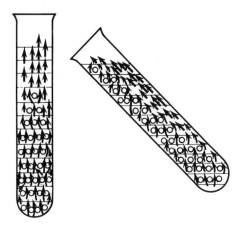

FIGURE 7-2   *Sedimentation in the angle head. Particles need not move the full length of the tube against the viscosity of the medium. (Courtesy Ivan Sorvall, Inc.)*

are available, such as a horizontal rotor for spinning the small tubes used in blood analysis.

The larger "high-speed" centrifuges exist in the greatest variety. Most modern units rotate at speeds up to about 20,000 rpm, developing centrifugal fields up to about 50,000 × g. As air friction generates considerable heat, most of these centrifuges are refrigerated. Because friction on the bearings is reduced to a minimum, the rotor would spin freely for some time after the motor is shut off if it were not for the magnetic brakes. Instruments now made by several manufacturers provide for continuous flow of liquid through the instrument while the rotor is turning at high speed. Sediment is collected in centrifuge tubes and the clarified liquid flows into a collecting ring and then into a separate container.

Preparative ultracentrifuges develop centrifugal fields up to about 400,000 × g. Refrigeration is provided. Air friction becomes so important at speeds up to 60,000 rpm that the rotor of this instrument turns in an evacuated chamber. The liquid to be centrifuged is placed in sealed tubes or inside a sealed rotor. Several different rotors are available, some of which have capacities of about a liter.

Analytical ultracentrifuges usually spin very small samples

of material, primarily for the purpose of determining sedimentation rate. The rotor must turn at very high speeds. Thus, compressed air bearings, magnetic bearings, or other novel bearings are used. Sedimentation in the sample tube can be viewed or photographed through a special illumination and optical system. Flashes of light are synchronized with the rotation, so in effect the rotation is stopped. The rotor, of course, operates in a vacuum under refrigeration. Some commercial models are available, but some of the analytical ultracentrifuges are built in the laboratories where they are used.

SEDIMENTATION   The rate at which particles will settle, or sediment, in a centrifugal field can be measured by means of an analytical ultracentrifuge. The information thus gained can be extremely useful in two different ways. First, this measurement provides information about the size, shape, density, or molecular weights of the particles. The molecular weight is an especially important thing to know. Analytical ultracentrifugation has become the standard method for finding molecular weights of proteins, nucleic acids, etc. Second, if the size and other characteristics of the particles are known, the length of time required to separate the particles from the suspending liquid can be computed, perhaps saving a good deal of trial-and-error work.

The analytical ultracentrifuge can be used to find the molecular weights of proteins, nucleic acids, and other large molecules. Similar techniques provide comparable information on the sizes of larger particles including viruses, ribosomes, and other particulates isolated from cells.

Two different methods are commonly used to find molecular weights. The first is the sedimentation velocity method. The rate at which the particles are settling through the suspending liquid is determined in the special rotors and optical systems of the analytical ultracentrifuge. In the second method, sedimentation equilibrium, the particles are allowed to settle as much as they will. At equilibrium, the downward, sedimenting force is balanced by a tendency to diffuse upward. In both methods, the basic principle is relatively straightforward, but corrections of various kinds make the computations somewhat complex. Sedimentation velocity measurements are quicker, but require information about the molecules from other sources. Sedimentation equilibrium measurements are easier to calculate, but the time required to reach equilibrium may be long. In practice, laboratories often run both methods on the same preparation of macromolecules.

SEDIMENTATION VELOCITY
Imagine a centrifuge tube containing a single protein molecule suspended in a liquid. In a strong centrifugal field, the molecule is subjected to a downward force

$$F\downarrow = m\alpha$$

where m is the mass of the particle and $\alpha = \omega^2 R$. A mole (N) of particles would have a mass equal to Nm, which is the molecular weight (M). Therefore,

$$F = M\omega^2 R$$

As the particles are accelerated downward they are opposed by a "frictional" force which is proportional to their velocity (v):

$$F\uparrow = fv$$

Here, f is a constant called the frictional coefficient. After the centrifuge has come up to full speed, the upward and downward forces establish themselves until $F\downarrow = F\uparrow$, at which time the particles settle at constant velocity. Then

$$M\omega^2 R = fv \quad \text{or} \quad M = f \times v/\omega^2 R$$

The value $\omega^2 R$ is a characteristic of the centrifuge and its speed of rotation; v can be measured. The frictional coefficient, f, must be evaluated, but will be constant for any single kind of molecule in a given situation. It (f) depends upon the shape of the molecule, the viscosity of the liquid, the buoyant force exerted by the liquid, the temperature, and other factors. Usually a value for f is computed from information gained in separate kinds of measurement, especially the ability of the molecules to diffuse. If other minor corrections are applied, we can change the symbol to K.

The value $v/\omega^2 R$ is the rate of sedimentation per unit of applied centrifugal field and has been given the symbol s. Depending on the size of the molecules, s ranges from about 1×10^{-13} sec to $100\text{–}200 \times 10^{-13}$ sec. The value 1×10^{-13} sec is taken as one Svedberg unit, S, named for The Svedberg, who developed much of the theory of sedimentation.

Thus, to recapitulate, s is determined by measurement of the velocity with which the particles migrate downward and from the characteristics of the centrifuge. The corrected frictional coefficient is a characteristic of the molecules and the liquid in which they are suspended; K is determined in separate kinds of measurements. The molecular weight can then be calculated: $M = Ks$.

Ribosomes and other particulate matter isolated from cells usually contain a number of molecules of nucleic acid, protein, etc., bound together in a discrete unit. Since it seems inappropriate to speak of the molecular weight of such particulates, such structures are usually characterized by giving the observed value for S. A typical small particle (RNA) might be called 4S (S = 4), another ribosomal fraction might be 70S (S = 70).

SEDIMENTATION
EQUILIBRIUM

Although it may take many hours to reach equilibrium, the sedimentation equilibrium technique offers the advantage that it is unnecessary to determine values for f separately. At equilibrium, the force downward, $F\downarrow$, is balanced by the upward diffusion of molecules. The net velocity, then, becomes zero, and f disappears. The downward force still depends on the mass of the particles. The upward diffusive force depends primarily on the concentration gradient within the centrifuge tube. Concentration will be highest at the bottom, lowest at the top. Measurements of concentration at two points within the tube, plus the information on $\omega^2 R$ and minor corrections, make it possible to compute the molecular weight.

Molecular weights determined by the two methods usually agree very well. For example, Svedberg found the values 68,000 and 70,000 for horse serum albumin.

DENSITY
GRADIENT
CENTRIFUGATION

One of the considerations in the estimation of particle weights by means of sedimentation is the difference between the density of the particle and the density of the suspending liquid. If the two densities happened to be equal, the particles would not settle at all. Density gradient centrifugation employs centrifuge tubes in which the suspending liquid varies in density. As would be expected, the density is greatest at the bottom of the tube, least at the top. If a suspension of particles is carefully layered on top, and then the tube is spun in the centrifuge, the particles will come to rest in a zone where the density of the liquid is the same as the density of the particles.

Density gradient tubes may be prepared in either of two ways. If one places a solution of cesium chloride in a centrifuge tube and applies a strong centrifugal field, the cesium chloride begins to settle and establishes its own gradient. If the other common ingredient, sucrose, is used, the gradient must be established in advance. This could be done crudely by pipetting into the tube a small amount of a concentrated sucrose solution—say 20 per cent sucrose. Very carefully, an equal amount of a slightly less concentrated solution would be pipetted on top of the concentrated solution. By adding successively less concentrated portions, one could build up several layers of sucrose, varying from 20 per cent at the bottom to 5 per cent at the top. The corresponding densities would be about 1.080 and 1.018. The gradient prepared in this way would be crude and somewhat irregular, but probably usable. For more uniform gradients, an ingenious device is available which withdraws liquid from two reservoirs—say 5 per cent and 20 per cent sucrose—and mixes the two liquids in the desired proportions. The gradients thus prepared vary in density continuously and smoothly from bottom to top.

Density gradient centrifugation might be used at rather low speeds to separate whole cells from more dense matter, or at higher speeds to separate subcellular particulates or macromolecules. Once the desired material has settled into the zone of corresponding density, it can be collected by carefully pipetting from the top. Even better, fractions can be collected by poking a hole in the bottom of the tube and catching separate small fractions (1–3 drops) in separate test tubes.

PROBLEMS   1. Calibrate a laboratory centrifuge. Measure the speed of rotation with a tachometer or stroboscope, being careful to keep the lid closed to prevent air pressure variations from changing the speed of rotation. Measure the radius of rotation and find the RCF at various levels in the tube, and at various settings of the speed control.

2. A partially ripe tomato should contain both green chloroplasts and orange-red chromoplasts, which may or may not

differ in density. It should be possible to prepare an extract of tomato cells and also to prepare a suitable sucrose density gradient. Do the two colors of plastids come to rest at the same level?

SELECTED REFERENCES

From the Bibliography:

Newman, David W., ed., *Instrumental Methods of Experimental Biology.* R. Trautman has a long, well-illustrated chapter.

Setlow, R. B. and E. C. Pollard, *Molecular Biophysics.* Covers ultracentrifugation well.

8

MICROSCOPY

If any instrument should be given credit for contributing more than any other to the development of biology as a science, that credit should go to the compound microscope. The history of detailed observation of living things parallels the development of magnification. Every improvement in the art of combining lenses made it possible to see smaller structural units. First the cells and later the minute structures inside cells were discovered. It is now possible to photograph biological structures even as small as single virus particles.

Although the microscope has contributed to biology, biology has contributed to microscopy as well. The need for better magnifying systems with which to examine parts of organisms has been the most important stimulus for the opticians.

The microscope was invented in the seventeenth century, and it did not take the original microscopists long to realize the importance of a number of factors, such as the quality of lenses, the way they were combined, and the type of illumination. By the end of the nineteenth century, the microscope had become a precision instrument. The theoretical and practical ability of the great German scientists of the time produced microscopes almost as good as any available today. The formulation of the cell theory and the accumulation of the wealth of information on the cellular structure of plants and animals occurred at this time.

The microscope permitted the biologist to see a great variety of cells, combinations of cells, and parts of cells. Each living thing is slightly different from any other living thing; some are composed of more-or-less standard parts, like mass-produced machinery, whereas others contain very special individual components. One of the tasks of those biologists who were able for the first time to examine cells in detail was to describe what they saw. Descriptions require words, mostly nouns and adjectives. Because of the great diversity of living things, a tremendous vocabulary of strictly biological terms arose. Any beginning biologist at that time faced the initial barrier of learning this vocabulary. To a certain extent this is still true, but unfortunately, there are some who feel that this vocabulary *is* biology and consider the terminology more important than the concepts the words were invented to describe.

At any rate, the importance of the microscope in the development of the science of biology cannot be minimized. The convenient models available now are used almost daily in every laboratory. The materials used by the experimental biologist are made up of cells which are too small to be seen by the unaided eye. The investigator interested in muscle contraction must consider changes in the appearance of muscle cells. Microorganisms are favorable experimental organisms, but must be examined routinely. Because living cells are below the limit of visibility, the compound microscope will always be one of the most useful tools to either the descriptive or the experimental biologist.

THE COMPOUND MICROSCOPE

The compound microscope is so called because it contains two sets of lenses: an objective lens, which produces an image, and an eyepiece or ocular, which further magnifies the image. The magnification available is the product of the magnifications produced by the separate lens units. The objective and ocular systems of the compound microscope are elaborate combinations of individual components fitted together according to a formula which is intended to provide the best possible view of the object being examined.

Microscopes differ in details of construction, partly because of competition among producers, but all are constructed according to a general pattern. Figure 8-1 shows a typical microscope, from which the major parts can be identified. The two lens systems, the objective (*a*) and the ocular (*b*), are at the opposite ends of a body tube (*c*). The object to be examined is placed on a glass slide on the stage (*d*) just below the objective lens. The object is illuminated from below through a hole in the stage, and the light may be focused by a substage condenser (*e*) and regulated by a diaphragm or other devices. The microscope is mounted on a strong supporting arm (*f*), which also serves as a handle. The body tube is moved upward or downward by a coarse (*g*) and a fine (*h*) focal adjustment in order to bring the object to focus at the eye point. Most microscopes have two, three, or four interchangeable objectives, all mounted on a revolving nosepiece (*j*). The slide and object may be moved about by a mechanical stage (*k*).

To the conservative oldtimer, some of the newer models hardly look like microscopes. Figure 8-2 shows one of these. The body tube contains a set of prisms arranged to form the image at the eyepiece, as in the conventional microscope.

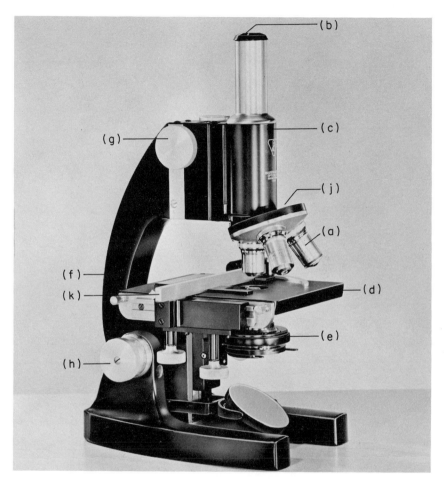

FIGURE 8-1   *A typical compound microscope. For explanation of labels, see text. (Courtesy Bausch and Lomb, Inc.)*

The prism system, however, permits the tilting of the eyepiece for greater comfort. In many of the new models, the prism system also divides the light between two ocular lenses so that both eyes can be used in viewing. Focusing is accomplished by raising and lowering the stage rather than the body tube. This arrangement is advantageous because the eye and head need not move upward and downward, but it may make the control of illumination from below slightly more difficult. Several of the newer models, including this one, even incorporate a third ocular tube to which a camera

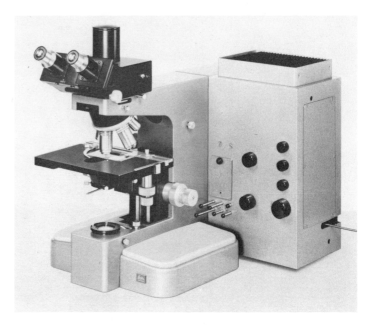

FIGURE 8-2   *A recent model, high-grade microscope. (Courtesy E. Leitz, Inc.)*

can be attached. The object can thus be located and examined and then photographed very conveniently.

The object might be illuminated with daylight from a window by means of a substage mirror, but nowadays artificial light is used more commonly. Light from a lamp can be directed toward the mirror or a small lamp can be placed beneath the condenser in place of the mirror. The microscope shown in Figure 8-2 is typical of the newer models, in that a source of illumination is built into the frame as an integral part of the instrument.

Figure 8-3 is a diagram of image formation in a compound microscope. The objective lens produces an image of the object at the level of the eyepiece. Further magnification occurs when the eyepiece, with the help of the lens of the eye itself, focuses the image on the retina. The eye interprets what it sees as if the ocular lens produced an enlarged image, or virtual image, at a distance of about 250 mm from the eye. Probably the ability to imagine the image at that distance requires some practice initially, but it comes quite naturally to anyone with a little experience.

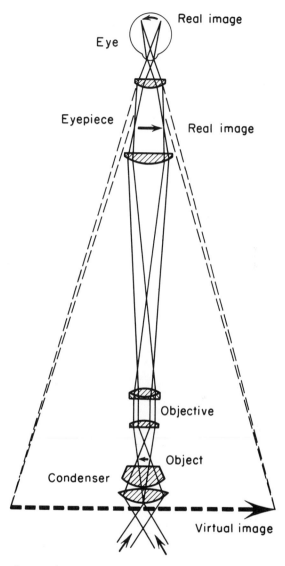

FIGURE 8-3 *Image formation in the compound microscope.*

OPTICAL THEORY All optical instruments, including the eye, depend upon light. Light moves as vibratory energy in straight lines outward from a source. In any one medium it moves with constant speed, but the speed is different in a different material. Any point on a moving beam can be pictured as a new source of

light from which new beams spread. The vibration occurs
in all the planes perpendicular to the direction of movement.
These statements will be elaborated in the following sections.

Visible light is but a small part of a much broader
spectrum of electromagnetic energy. We commonly think of
this electromagnetic energy as vibratory in nature, and its
movement can be pictured as a wave (as shown in Fig. 8-4).
The wavelength is the distance from any point on one wave
to the same point on the next wave, as indicated in the
diagram. Another feature of vibratory energy is a frequency,
which is the number of vibrations in a unit of time. In the
diagram, curve A has a low frequency, while curve B has
a higher frequency. The speed of light (c) is constant
(3×10^{10} cm/sec in a vacuum), and the frequency and the
wavelength multiplied by each other must equal this constant.
In other words, the speed of light becomes a proportionality
constant, and we arrive at this equation:

$$\lambda = \frac{c}{\nu} \quad \text{or} \quad \lambda \nu = c$$

Lambda (λ) is the symbol commonly used for wavelength
in centimeters or in fractions or multiples of centimeters.
Frequency is given the symbol nu (ν), and is expressed in
reciprocal seconds (sec^{-1}), or what amounts to vibrations
per second. Another useful way of describing the frequency
is in terms of a wave number, that is, the number of vibrations
in a certain unit of length. In the electromagnetic spectrum
we find a continuous variation in wavelength and frequency
from one end of the spectrum to the other. At one extreme,
wavelengths are very long, and frequencies are very low.
At the opposite end, wavelengths are very short, and frequencies are high. The visible region of the spectrum lies
near the middle of the much broader spectrum. There is
nothing unique about visible light; it merely corresponds to
the ability of our eyes to perceive light. The red end of the
visible spectrum consists of the longer wavelengths and lower
frequencies, while the blue-violet end has shorter wavelengths
and higher frequencies. Immediately beyond the visible
spectrum and continuous with it are the infrared and the
ultraviolet. It is difficult to draw a line between the red,
visible light and the infrared, but usually this line falls around
a wavelength of 700 nm. The average human eye can see

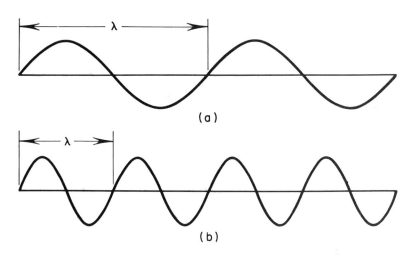

FIGURE 8-4   *Wave representation of radiant energy. In curve* A, *the wavelength* (λ) *is twice as great as in curve* B; *in curve* B *the frequency is twice as great as in curve* A.

wavelengths as short as about 400 nm; anything below this limit lies in the ultraviolet.

REFRACTION   When a light beam passes from one medium into another in which it travels more rapidly or more slowly, the direction of movement is changed. The moving beam can be imagined as a progression of "wavefronts" moving outward from a distant source, each front one wavelength ahead of the next. Near the source these fronts will be spheres, but, as the distance from the source increases, any small segment of the sphere behaves as a plane. If such a beam of light, moving through air, approaches the surface of a piece of glass— through which it moves more slowly—either of two events might occur. The beam might be reflected with an angle of reflection equal to the angle of incidence, or the light beam might enter the glass. However, since it moves more slowly in the glass, any oblique beam will change direction slightly upon entering the glass. The angle of bending depends upon the ratio of the speeds in the two media. Figure 8-5 shows a train of wavefronts approaching an air-glass surface. The angle of reflection is equal to the angle of incidence, but the direction through the glass is different. This bending, resulting

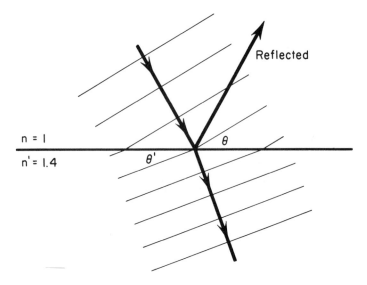

n = 1

n' = 1.4

FIGURE 8-5   *Refraction at an air-glass interface.*

from the entry into a different medium, is called refraction.

The angles θ and θ', which the wavefronts make with the surface, depend upon the speed of light in the two media according to the relationship

$$\frac{\sin \theta}{\sin \theta'} = \frac{v_1}{v_2} = \frac{n_2}{n_1}$$

where v_1 and v_2 are the speeds in the two media, and n_2 and n_1 are the respective refractive indexes.

The refractive index is a ratio comparing the speed of light in any medium with that in a vacuum. Because the refractive index of air is only very slightly greater than unity, for practical purposes, glass and other materials are commonly referred to air as a standard.

Although it cannot be seen from Fig. 8-5, the frequency of vibration of the light beam remains constant, the speed of propagation changing because the wavelength changes. The degree of change in wavelength is dependent upon the wavelength itself. For example, shorter wavelengths are retarded more upon passing from air to glass. For this reason, the refractive index of a material is expressed relative to a certain wavelength of light, usually the sodium line at $5893\mathring{A}$ (589.3 nm).

When light passes from glass to air, the same kind of bending will occur, but in the opposite direction. Thus a pair of parallel surfaces causes a jog in the light path, but it is not apparent because the entering and emerging beams move in the same direction. A lens, however, has opposite surfaces inclined at angles to each other, so that the emerging beam may have quite a different direction from the entering beam. In a microscope, every plane in the light path where light passes from one medium to any other must be given careful consideration.

DIFFRACTION   According to Huygen's principle, every point on an advancing wavefront acts as a new source of light. This leads to an apparent bending of light around corners, known as diffraction.

If two parallel wavetrains happen to be vibrating in the same direction, their amplitudes are additive, and the two waves are said to be in phase. If the two waves vibrate in opposite directions, they cancel each other, and interference results (see Fig. 8-6).

Imagine a beam of light of one wavelength approaching a barrier with two parallel slits, as in Fig. 8-7. Each slit acts as a new source of light. Point 0 on the screen is the same distance from *a* as from *b*, so light arriving from *a* and from *b* are in phase and reinforce each other, producing a bright spot on the screen. Point 1 on the screen is one wavelength farther from *a* than from *b*, but again the two beams are in phase. Midway between 0 and 1, light from *a* and from *b* are exactly out of phase, with the result that they cancel each other, giving a dark spot on the screen. The pattern seen on the screen, then, is a series of bright lines at 0, 1, 2, etc., and also on the other side of 0, as at 2'.

If another color of light with a longer wavelength is used, similar lines appear, but the distance between 0 and 1, and between all other lines, is greater. If white light, with its mixture of all wavelengths, is used, the result is a dispersed spectrum with the blue end nearer the center (point 0) and the red end more remote. Each point on the screen (1, 2, etc.) becomes a complete spectrum.

A diffraction grating consists of a large number of parallel lines drawn on a surface. As long as the lines are almost perfectly parallel and the spacing of the lines is uniform, the large number of lines produces a diffraction pattern like

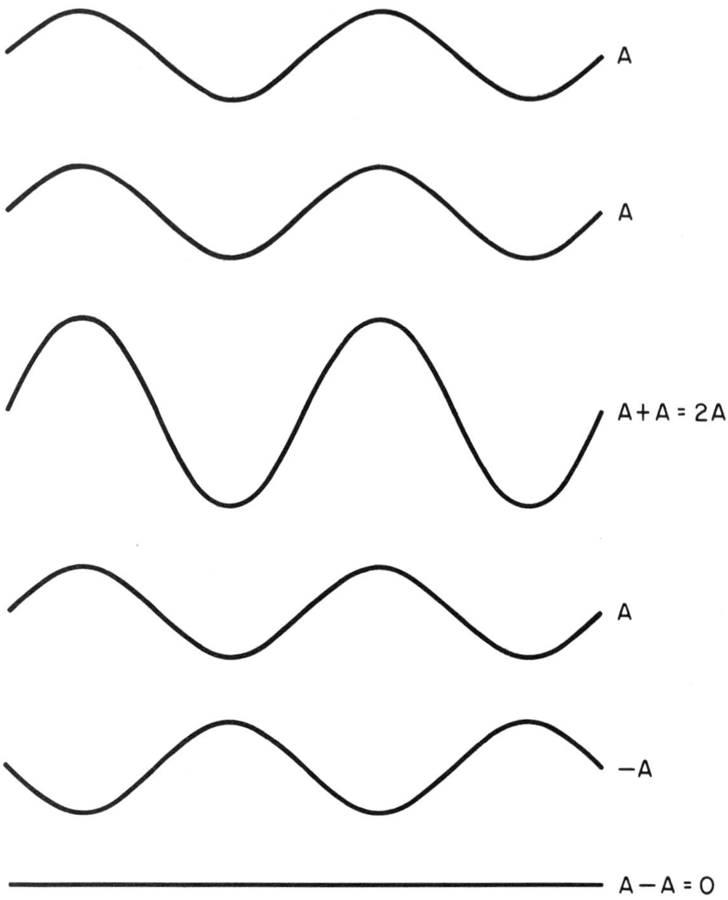

FIGURE 8-6   *Reinforcement and interference. In the upper set of curves, the waves are in phase, and their amplitudes are additive. In the lower set of curves, the waves are exactly out of phase, so they cancel each other.*

that resulting from a single pair of slits. The grating disperses a beam of white light into a number of complete spectra, labeled first order, second order, and so on. The preparation of diffration gratings for use in optical instruments is an elaborate process. Lines—actually grooves—are engraved on a glass blank with an optically flat surface by means of a very precise "ruling engine." Because the master grating is extremely expensive to make, practical and useful "replicas"

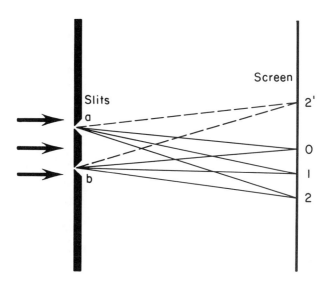

FIGURE 8-7   *Diffraction pattern produced by two slits. For explanation see text.*

are molded in a plastic material, using the master grating as a pattern or model. The replica grating may transmit light or it may be silvered for use as a reflection grating.

Diffraction also occurs in other materials with regions of difference in light transmission or in refractive index, but irregular shapes and spacings are much harder to analyze.

POLARIZATION   Light ordinarily vibrates in all possible planes. Imagine light descending vertically from the sun. It vibrates not only in the north-south plane but also in the east-west plane and at every point of the compass. In certain situations, as by reflection at a certain angle or by passage through special materials, the vibration is restricted to a single plane. Such light is polarized. The ability to polarize a beam, to permit the passage of a beam of polarized light, or to rotate the plane of vibrations all depend upon the nature of a material, either the molecular structure or the arrangement of molecules in some larger unit. In ordinary microscopy, polarization is of no consequence, but the special techniques of polarization microscopy, described in a later section, can give valuable clues to the structure of materials.

MAGNIFICATION AND RESOLUTION   The magnification of a microscope is stated by the manufacturer. An objective with "43 $\times$" printed on the side might be used with an eyepiece labeled "10 $\times$." The magnification by the objective lens refers to the ratio of the size of the image produced to the size of the object at a specified body tube length. If the body tube is made longer than the customary 160 mm, the magnification by the objective is increased. The magnification by the eyepiece is calculated, assuming the virtual image at 10 in. or 250 mm from the eye point. The total magnification, using 160 mm tube length and virtual image at 250 mm, is the product of the two stated figures, or 430 $\times$ in this example.

Magnification is essential in order that images of individual spots on the object will fall on the retina of the eye sufficiently far apart from each other to excite nonadjacent cells. If the magnification is inadequate, two separate spots will strike the same or closely adjacent retinal cells and will be interpreted as a single spot. In addition, the object to be examined may occupy a very small portion of the field of view and enlarging it leads to a more favorable view. If the magnification is too great the object may appear much larger than the field of view. The "zoom" microscope offers continuously variable magnification and can be adjusted to provide the most convenience in viewing.

By building a microscope with a very long body tube, any measure of magnification could be achieved. Given a bright light source, a very large image of a bacterial cell could be cast upon the moon. But magnification alone is not the most important quality in microscopy. The ability to distinguish or resolve small objects is much more important. Further magnification does not necessarily improve the ability to see details; only an increase in resolving power can do this.

The limit of resolution of any optical system is defined as the ability to separate adjacent objects. Under a microscope two objects or lines can be seen as separate and distinct. As two lines are moved closer together, they eventually seem to merge into a single line. The distance between the lines just before they can no longer be distinguished from each other is the limit of resolution. This limit, also called resolving power, for the human eye is about 0.1 mm. Since this is about the maximum size of ordinary cells, the microscope becomes very necessary in biology.

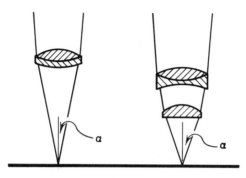

FIGURE 8-8   *A pair of objective lenses, to show the angle through which a lens can gather light. Here α is the half-angle. Angular aperture = 2α.*

The resolving power of a lens system depends upon the angle through which it can gather light. Figure 8-8 shows a pair of objective lens systems. One has a long working distance and gathers light through a small angle. The other gathers light over a wider angle and would be expected to resolve smaller objects. The angle has been called the angular aperture (A. A.), and its size can vary from a few degrees to a theoretical maximum of 180°. The refractive indexes of the lens, of the slide and cover glass, and of the substance between the cover glass and the lens all influence the resolving power, however, so that angular aperture alone is not an adequate expression of the ability to resolve. Another expression, the numerical aperture (N. A.), more nearly indicates the resolving ability.

N. A. $= n \times \sin \frac{1}{2}$ A. A.

n being the refractive index of the least refractive material between the object and the objective lens. The greater the value of N. A., the better the resolution. An objective lens with a working distance of 16 mm, labeled 10 $\times$ by the manufacturer, has a numerical aperture of about 0.25. With other objectives the N. A. values range upward to a practical limit of about 1.4.

Ernst Abbe, probably the greatest of the German opticians, proposed the value N. A. and developed the theory of resolving power of microscopes. Perhaps at low magnification the image seen in the microscope depends upon

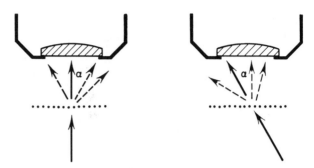

FIGURE 8-9   *In the figures, α is the angle between the axial ray and the first diffracted ray. On the left, axial lighting is used and the lens would capture the blue, but not the red diffracted light. Oblique illumination (right) increases the effective angle through which the lens gathers light.*

transmission and absorption of light by the object. The examination and interpretation of very small units, however, probably depends upon diffraction phenomena. Each small object is accompanied by a series of interference bands which are responsible for the contrast observed. If the interference bands of two objects overlap, the two objects appear as one and cannot be resolved. Abbe reasoned that the limits of resolution could be found by examining a diffraction grating. If the lines of the grating are to be seen, the objective lens must include the direct or axial ray plus at least the first diffracted ray (Fig. 8-9). The angle (α) depends upon the wavelength (λ), upon the distance (d) between lines on the grating, and upon the refractive index of the medium, or $n \sin \alpha = \lambda/d$. However, $n \sin \alpha =$ N. A. so that $d = \lambda/$N. A. This is the smallest value of d resolvable by an objective. If oblique illumination is used, as from an Abbe condenser, this limit is reduced to about half, or $d = \lambda/2$ N. A. This limit is unattainable for several geometric reasons; the theoretical limit of resolution is usually given as 1.2 $\lambda/2$ N. A. If an object is illuminated with blue light ($\lambda = 400$ nm) and examined with a lens system (N. A. $= 1.40$), the limit of resolution is

$$d = \frac{1.2 \times 400}{2 \times 1.40} = 170 \text{ nm}$$

or about half the wavelength of the light used.

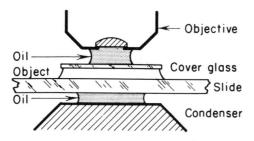

FIGURE 8-10   *Oil immersion microscopy.*

An alternate method for calculating the resolution limits starts with point sources of light instead of lines on a grating. As two bright spots at the level of the slide are brought closer together, it eventually becomes impossible to resolve them as separate sources of light. The calculation involves diffraction and interference phenomena, as does Abbe's method. The limits of resolution calculated by the two methods are about the same. Probably Abbe's technique more nearly corresponds to the events of image formation when the microscope is used in a normal manner.

Since the numerical aperture is limited by the refractive index of the least refractive medium in the light path, high-resolution microscopy uses oil-immersion lenses. A drop of a transparent "oil" with a refractive index of about 1.5 is placed between the cover glass and the objective lens and usually also between the condenser and the slide (see Fig. 8-10). The material to be examined is mounted between the slide and the cover glass in another medium of high refractive index. The numerical aperture is thus increased, reducing the limit of resolution.

An objective lens is moved upward or downward until the object is in focus on the retina. A slight deviation upward or downward produces an inferior image, but the eye might not notice the imperfections. Further deviation, however, produces an image which is obviously out of focus. The thickness of the object which seems to be in focus at any one time is known as the depth of field. The greater the resolving power, the thinner this depth will be. If great detail is to be seen, high numerical aperture is required, but visibility of details is achieved with a sacrifice of depth of field. Usually an object is examined first with a low power objective (greater depth) and later with the high power objective. The narrow depth of field in high power objectives is not entirely a disadvantage because it permits "optical sections" of cells. If the center plane of a cell is in focus, both the top and bottom will be invisible and thus will not confuse the image of that center plane. The net result is the same as taking a very thin slice out of the center of the cell. By successive examinations of various optical sections within the cell, a fairly complete three-dimensional picture can be built up.

ABERRATIONS   High quality optical components for microscopes must incorporate corrections for certain "aberrations" if the image is to represent the object. Otherwise several aberrations inherent in optical systems can be extremely annoying.

SPHERICAL
ABERRATION   Lens elements with spherical curvature do not focus at a point because the edges produce relatively greater refraction than the center. An object cannot be brought to focus at a point but rather is spread over some distance along the optical axis. This spherical aberration could be corrected by using lenses with nonspherical curvature, but microscope components are so small that the sphere is about the only curvature that can be made practically. More commonly, spherical aberration is corrected by combining lens elements in such a manner that one lens compensates for the spherical aberration of another.

CHROMATIC
ABERRATION   Because refraction depends on the wavelength of the light, a single lens does not focus all colors at the same point. The object seems to be surrounded by halos of different colors. This color discrepancy is called chromatic aberration. It can be corrected by using combinations of lens elements of different refractive index. Achromatic objectives composed of different kinds of glass achieve adequate correction for ordinary purposes. High-resolution microscopy demands more correction, however, and apochromatic objectives made up of several elements of glass and the mineral fluorite are used. Achromatic and apochromatic objectives correct for spherical as well as chromatic aberrations.

Corrections for the other aberrations (astigmatism, coma, curvature field) are of little concern to the microscopist since the manufacturers have nearly eliminated these problems. Microscopes corrected for spherical and chromatic aberration are magnificent instruments, but they must be used properly. For example, correction for spherical aberration was designed for use with a cover glass 0.18 mm thick, and any other thickness introduces spherical aberrations.

ILLUMINATION   High resolution microscopy is impossible without adequate illumination. The modern microscope incorporates a substage condenser to focus a beam of light on the object. If the light is not properly directed, the effect on the resolution is just as bad as if the object were out of focus.

Several methods of illuminating objects have been described and are in use. Each system includes a light source and lenses and diaphragms. Setting up the illumination system requires painstaking attention to detail. The best source of instructions is the manual provided by the manufacturer of the microscope. It is important to rely on this manual, because the new microscopes often have novel and unique systems which achieve the same purpose as the older systems. The new systems have been carefully designed, and unless properly used the image will be inferior.

PREPARATION
OF MATERIALS   Some biological materials can be placed on a slide with water, covered by a cover glass, and examined immediately. Many others, however, are too thick to transmit light or there is inadequate contrast between parts. The biological specialty known as microtechnique has evolved as a result of the need to prepare materials for examination.

Water mounts are satisfactory for some purposes, but the water tends to evaporate. If a specimen must be examined for a longer period of time, mounting in glycerol gives better results. If living materials must be kept alive during examination, either water or glycerol is satisfactory, and certain non-poisonous dyes can be added. Neutral red, methylene blue, and janus green are "nontoxic" and tend to collect mainly in certain parts of cells, increasing the contrast between these parts and the rest of the cell. If the material is too large or too thick, slices can be made freehand with a razor blade. With practice, a steady hand can cut slices only one or two cells thick.

Prepared slides or permanent mounts are made when more than a preliminary examination is required. The techniques are complex and variable, are frequently long and tedious, but usually involve four major steps: fixation, embedding, sectioning, and staining and mounting. The order of the steps may vary somewhat. Fixation is a method of killing the cells. Ideally, the living material is killed rapidly and in such a way that the various structures do not become disarranged with respect to each other. We should hope that the nucleus of a dead cell on a prepared slide looks about the same as the living nucleus. Obviously changes do occur, but certain fixation procedures apparently bring about a minimum of change.

Before a fixed and softened material can be sliced into thin sections it must be provided with some support. Materials prepared for light microscopy are commonly embedded in paraffin. Paraffin melts at a fairly low temperature, so the biological material ordinarily need not be damaged by a high temperature treatment. More recently several synthetic plastic materials have been used. Embedding for electron microscopy employs an epoxy resin or a methacrylate polymer —similar to Lucite and Plexiglas—which is prepared from small monomer molecules during the embedding process.

Slices or "sections" of a block of embedded tissue are cut on a microtome, an example of which is shown in Fig. 8-11. The block of material is held in a clamp and moved mechanically over a very sharp blade where a thin slice is cut off. On the next stroke, the embedded tissue is advanced toward the blade by a distance equal to the desired thickness of the tissue slice. Each stroke thus advances the tissue, usually by means of a lead screw, and then cuts off the amount of the advance. The sections of tissue, surrounded by the embedding material, slide off the knife, one after another, in a more-or-less continuous ribbon. If a series of sections is prepared in this way, each slice can be studied separately, and the whole series allows the biologist to assemble a three-dimensional picture. The microtome is a rugged but precise instrument, permitting the slicing of sections thinner than a single cell.

Permanent prepared slides are made from sections cut on the microtome. The section of biological material is attached to the slide, and the embedding material is dissolved away. Further treatment involves soaking in one or more stains

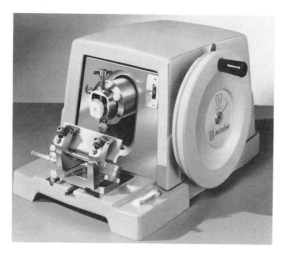

FIGURE 8-11   *A rotary microtome. (Courtesy American Optical Company.)*

or dyes, chemical treatment to make the staining permanent (at least in certain parts of the cell), and washing to remove excess dye. Dyes are selected on the basis of their affinity for certain chemical substances in the cells. For example, one dye might be attracted only to lignin and thus would collect in the cell walls of plant cells, staining them, say, red. A second staining treatment might stain cytoplasmic proteins green. The finished slide would show individual cells with red cell walls and green cytoplasm, which certainly is more contrast than the original material furnished. Once the staining is completed and chemical treatment has withdrawn all water, a cover glass is cemented over the biological material with a natural or synthetic cement of high refractive index. The whole preparation might require several hours or days, but slides prepared in this manner last for years and can be examined under the microscope at any time. If the slide has been well made, there will be a minimum of disarrangement and swelling or shrinkage of the parts of cells.

THE PHASE CONTRAST MICROSCOPE   Some biological materials and some parts of cells, especially when alive, offer so little contrast in density, refractive index, or color that they are difficult to study with an ordinary microscope. Some of these can be studied more effectively through the techniques of phase microscopy.

In ordinary "bright field" microscopy, high resolution depends upon diffraction of light and the fringes that appear on the edges of structures. The diffracted light must be caught by the objective lens, and the image is formed from the interference pattern of the direct or axial light and the diffracted light. If there is sufficient contrast (difference in absorption, reflection, or refraction) at these edges, the image is sharp and objects are easily recognizable. Sometimes when the contrast is high, the bright or dark interference bands around small objects are so obvious that they become annoying. If the contrast is insufficient, the diffracted rays will be weak, the direct rays will seem too bright, and the object will be difficult or impossible to see. Phase contrast is basically a means of balancing the direct and the diffracted rays to increase the visibility of the interference fringes.

As a rule, the direct rays and the diffracted rays travel to the image level by paths of different length and therefore arrive out of phase with respect to each other. One of the features of the phase contrast microscope is a modification of these phase relationships to provide a more favorable image. In addition, the direct rays are commonly reduced in intensity so that diffracted rays contribute relatively more to the image. The diagram in Fig. 8-12 shows one common means of achieving this balance. The ring-shaped opening (annulus) in the condenser causes the illumination of the object by a hollow cone of light. The undiffracted rays pass through the objective to the "phase plate" where they are modified. Diffracted rays pass through other parts of the objective lens and phase plate. Various types of plates are available to produce bright images against a dark background or dark images against a light background. Boundaries within the object that differ only slightly in refractive index might not be visible in the ordinary microscope but are seen easily with phase contrast.

Phase contrast microscopy requires a good deal of experience and study. Very faint boundaries or edges become readily visible, and it is too easy to see structures that are not real. A wrong choice of the type of phase modifying plate could give a completely erroneous impression of the object. Skilled users, however, have seen and photographed parts of living cells that would be completely invisible without the phase contrast microscope.

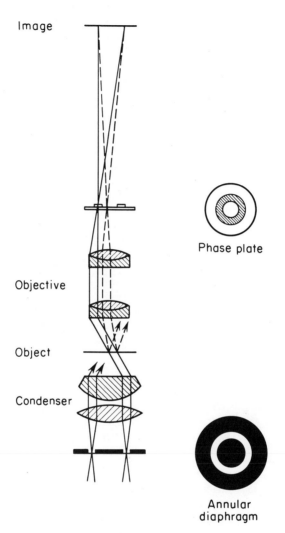

Image

Phase plate

Objective

Object

Condenser

Annular
diaphragm

FIGURE 8-12   *Phase contrast optics. The ocular is not shown. Diffracted rays (dashed lines) take a different path through the instrument than direct rays (solid lines). The phase plate modifies and balances the light from the two sources.*

THE POLARIZING The use of the polarizing microscope yields information
MICROSCOPE about materials with an ordered internal structure. Chemists use this microscope to examine crystals because the crystal contains atoms or molecules arranged in a definite order. Certain biological materials seem to have similar definite

arrangements, and information about this fine structure can be obtained by studying the materials under polarized light.

In the polarizing microscope, light is supplied through a disk of polarizing material, such as high-grade "Polaroid," so that the specimen is illuminated by light which vibrates in only one plane. The disk of polarizing material, known as the polarizer, is mounted in or adjacent to the condenser in such a manner that it can be rotated. A similar disk of polarizing material, called the analyzer, is mounted behind the objective or in the eyepiece and also can be rotated. Scales marked in degrees tell the direction of orientation of the polarizers and analyzers. If the two are parallel to each other, the polarized light passes through the analyzer easily, and the field appears bright (with no specimen on the stage). If the analyzer is rotated 90°, it prevents the passage of the polarized light, and the field will be dark. This is the condition known as "crossed polarizers." Polarization microscopes are also fitted with a rotating stage so that the object can be rotated within the polarized illuminating beam.

EFFECTS OF MATERIALS ON THE POLARIZED BEAM Some kinds of materials have no effect upon polarized light regardless of the direction in which the light passes through. The polarizing microscope will behave the same as if the material were not in the light path. Such materials are optically isotropic. Other materials, notably crystals and colloidal aggregates, show a different behavior depending on the direction in which the light beam passes through them. Such materials exhibit one or more of several forms of optical anisotropy.

Most optically anisotropic materials differ in refractive index in different directions, which means that the polarized light passes through the material more rapidly in one direction than in others. Such materials are doubly refractive or birefringent. The effect would be apparent in a material such as that shown in Fig. 8-13. Rods of one substance, with a characteristic refractive index, are aligned parallel to each other in a medium of different refractive index. The polarized light is propagated in the direction parallel to the fibers with a velocity different from the velocity across the fibers. If this birefringence is a property of the material itself and not of the medium in which it is mounted under the microscope, the material exhibits intrinsic birefringence. A similar

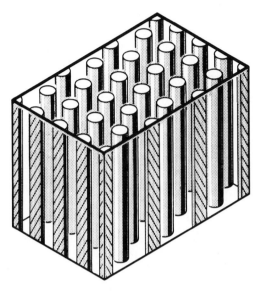

FIGURE 8-13   *Hypothetical birefringent structure. Polarized light passes through such a structure more rapidly in some directions than in others.*

effect occurs when particles with one refractive index are oriented in a medium of different refractive index, but this effect—form birefringence—might disappear if the medium is changed. Components of protoplasm seem to show both these effects under certain conditions.

Dichroism is dependent upon differences in absorption of polarized light passing through in different directions. Certain colors of light are absorbed when the material is oriented in a particular way, while another pattern of absorption appears if the material is rotated 90°. Thus the material will change color as it is rotated.

The analysis of materials with the polarizing microscope depends upon changes occurring when the polarizer, the material, or the analyzer are rotated. The observer may see alternations of light and dark fields, changes in color, or a series of color fringes reminiscent of spectra. Interpretations require experience, and only relatively few biologists have used the polarizing microscope. There is a body of literature including studies on protoplasm, however, and some of our knowledge of the orientation of materials in protoplasm has resulted from such studies.

Another fairly simple effect of some materials on the beam of polarized light is a rotation of the plane of polarization. If the polarizer of the microscope is set so that the light is vibrating in the "twelve o'clock to six o'clock" plane, the material on the stage will rotate this plane to the right or left by an amount which depends on the nature of the material and its thickness. The degree of rotation can be determined by rotating the analyzer until the field looks bright. Ordinarily this effect is small, and the polarizing microscope would not be used to measure it. Instead, we would use an instrument called a polarimeter, which is similar in principle but has provision for a much thicker layer of the material to be examined.

THE ELECTRON MICROSCOPE

The electron microscope employs a beam of electrons instead of a beam of light. These electrons, which are produced by a heated filament, can be focused into a beam by an electrostatic or a magnetic field. The electrons behave as if they had a frequency and a wavelength, but this wavelength is much shorter than the wavelengths of visible light. The "optical" parts of the electron microscope are analogous to those in the light microscope, consisting of an electrostatic or magnetic "objective lens" and a similar "projector lens." The electron beam passes through the object, where electrons are either transmitted or scattered in various directions, depending on the nature of the material in the object. The transmitted electrons are brought to focus on a photographic plate in a pattern corresponding to regions of high transmission or high scattering in the object. Since the wavelength of the electrons is short, the resolution is greater than that available in the light microscope.

Biological materials offer difficulties in electron microscopy, but the solution of these problems has permitted pictures showing exceedingly fine details of structure. Most recent biology books contain excellent examples. Even though these show great detail, much is lost in the printing process; the original photographs are truly magnificent.

The electron beam must operate in a vacuum, which means that the biological material must be dry (and therefore dead). Exceedingly thin layers of material must be used, and usually, since biological materials are relatively ineffective in electron scattering, atoms of metal are added to increase contrast.

Earlier electronmicrographs were prepared by drying a thin film of biological material and then "shadowing" witl metal atoms. In a vacuum chamber the metal atoms are "sputtered" off a heated coil from a position above and to the side of the biological material. The metal atoms form a thick layer on one side of any raised places in the biological material and a thin "shadow" behind these spots. The resulting electronmicrograph provides a three-dimensional effect more or less like an aerial photograph, with alternations in light and shadow. This technique is still used, but other methods have become available.

More recently it has been possible to photograph internal details of biological structures. The tissue to be examined is chopped and then fixed in a solution of osmic acid (osmium tetroxide). The cells are killed, and, at the same time, metallic osmium atoms are deposited in certain parts of the cells. The fixed material is embedded in plastic and sectioned on a special microtome. The slices produced must be thinner (0.02μ, for example) than those used for light microscopy. These thin slices are supported on a screen analogous to the glass slide and placed in the electron microscope. Direct viewing is possible if the electron beam is focused on a fluorescent screen, or a photograph can be made.

A technique called negative staining has come into use recently, particularly for visualizing such macromolecules as proteins and nucleic acids. The biological materials are embedded in a matrix of electron-dense material. The organic molecules then show up as relatively clear areas against a dark background. Molecules and fibers as small as 2 nm in diameter are often strikingly visible in negatively stained preparations.

The theoretical limit of resolution of the electron microscope is in the vicinity of 2×10^{-10} cm, compared to a theoretical limit for the light microscope of about 2×10^{-5} cm. This calculation, however, assumes an electron "lens" of high N. A. In practice, magnetic and electrostatic lenses have large aberrations, and only low values of N. A. can be used. The practical limit of resolution for the electron microscope is about 1 nm, or 10^{-7} cm. This limit is still a considerable improvement over that found in the light microscope.

The interpretation of electronmicrographs requires some caution. The biological material has been subjected to a variety of treatments and of course is no longer alive. It is entirely possible that fixing, staining with metal, and sectioning have introduced "artifacts," or apparent features which are not real. The tissue will have been sliced in many different planes, and the selection of a particular spot to photograph depends on the microscopist's idea of what the preparation ought to look like. Nonetheless, so many fine pictures showing exquisite detail have appeared recently that the electron microscope has had a profound influence upon biological research.

PROBLEMS   1. Using the instructions provided by the manufacturer, set up a microscope and illuminator in such a manner as to achieve optimum illumination (critical, Koehler, or other system, depending on the instruments). Then examine a good prepared slide which shows fine details, and compare the view with what is seen when the light is not properly focused.

2. Electron micrographs of many biological materials can be found in modern textbooks, journals, etc. It is more difficult to find light microscope pictures of the same structures which would be useful in demonstrating the contrast between the resolving powers of the two kinds of microscopes. A search through the literature may turn up pairs of pictures which would illustrate this difference, or you may be able to make light microscope pictures to match some published electron micrographs.

SELECTED REFERENCES   Bradbury, S., *The Evolution of the Microscope*. New York: Pergamon Press, 1967. A thoughtful history placing the microscope in its proper position.

Clark, George L., ed., *The Encyclopedia of Microscopy*. New York: Reinhold Publishing Corporation, 1961. Complete coverage of all the various topics of microscopy.

Haggis, G. H., *The Electron Microscope in Molecular Biology*. New York: John Wiley & Sons, Inc., 1967.

Jones, Ruth McClung, *Basic Microscopic Techniques*. Chicago: University of Chicago Press, 1966.

Martin, L. C., *The Theory of the Microscope*. New York: American Elsevier Publishing Company, 1966.

The colorimetric procedure, one of the most common methods used in analytical chemistry today, finds application in biology also. The method depends upon those physical principles which are related to the color of various substances. In its simplest form, colorimetry measures the amount of material by measuring the intensity of its color. The greater the concentration, the more highly colored the solution. An extension and refinement of the technique is commonly given the term spectrophotometry. Spectrophotometry also can be used to determine concentrations, but has the added advantage that it can be used to identify materials and measure rates of reactions. Any material that has color, or more properly, any material that absorbs radiant energy in the visible region, in the ultraviolet, or in the infrared, is adaptable to measurement by this procedure.

GENERAL CONSIDERATIONS Measurement of the color of materials depends upon the nature of light itself. The wave nature of light was described in Chapter 8. The conception of light as wave motion, however, does not entirely and adequately describe its behavior. Although seemingly incongruous at first, it is now relatively easy to accept a different nature for light. It can be shown mathematically that when it interacts with matter, light behaves as if it were made up of corpuscles or packets of energy. These packets of energy are commonly known as quanta (singular: quantum) or as photons. A quantum is simply a certain amount of energy, whose energy equivalent could be measured in calories, electron-volts, or ergs. The size of the quantum increases as the frequency increases. The extremely long wavelength and low frequency radio waves contain relatively little energy per quantum, while gamma radiation, with its short wavelength and very high frequency, contains a great deal of energy per quantum. Planck was able to show that the relationship is in strict proportion and was able to arrive at a value for the proportionality constant (h) of the equation

$$E = h\nu \tag{9-1}$$

If we know the frequency, we can use Planck's constant, approximately 6×10^{-27} erg-sec, and arrive at the energy (E) in ergs per quantum. Since a single quantum is an extremely small amount of energy, we usually deal in terms of N (Avogadro's number) quanta. By way of illustration it can be pointed out that N quanta of red light are equivalent to about 40 kilocalories of energy, while N quanta of blue light are equivalent to about 70 kilocalories.

Colorimetry and spectrophotometry depend upon the interaction between light and matter. Almost every material will absorb radiant energy of some wavelengths. If these wavelengths happen to be within the visible region, we say that the material has color because the eye sees only those parts of the spectrum which are not absorbed but are reflected or transmitted. If the absorbed wavelengths happened to be in the infrared or ultraviolet, and relatively little of the visible light were absorbed, the material would be white or colorless. If nearly all the visible light were absorbed, we would call the material black.

When a molecule absorbs light energy it must absorb one and only one full quantum. It cannot absorb parts of quanta or several at a time. When the atom or molecule accepts this bundle of energy, so that it contains more than the usual amount of energy, it goes into an "excited state." The most likely explanation on the basis of atomic structure is that one of the orbital electrons is lifted from a stable configuration into a less probable, higher energy state. The amount of energy required for this transformation is exactly the amount in the particular quantum. In any kind of atom or molecule there may be several possible ways in which the different electrons can be displaced. Each of these would require its own characteristic amount of energy. Therefore the atom or molecule could absorb several different wavelengths of radiant energy. If we examine the material by means of the spectroscope we see dark lines wherever light is absorbed; the pattern of these dark lines is characteristic of the material being examined.

The absorption of light depends upon the displacement of electrons, and the phenomenon should also occur in reverse. If a molecule is heated to a sufficiently high temperature, electrons can be displaced. As they fall back to their normal levels they emit the excess energy as light. If we examine this light with a spectroscope we find that the bright lines, or emission lines, correspond at various regions in the spectrum with the dark absorption lines in our previous spectrum.

The molecule in the excited state is unstable, and the energy must be dissipated in a small fraction of a second.

There are several possible fates for this excitation energy. Far more common than any of the others is the conversion of this energy into heat, which is transferred to other atoms and molecules in the vicinity. The displaced electron falls back to its normal energy level. In some cases, the energy may be used to drive a chemical reaction. The whole field of photochemistry is founded upon this ability. The absorbing molecule itself may be changed in the photochemical reaction or it may serve as a sensitizer for some other reaction. A third possible fate for the quantum of absorbed energy is its re-emission as light in the process of fluorescence. Since, during the lifetime of the excited state, a small fraction of a second, a portion of the energy will be dissipated as heat, the quantum re-emitted must be smaller and therefore of a longer wavelength than the quantum absorbed. In certain kinds of molecules a fourth fate is possible. The excited molecule may change from the unstable excited state into a longer-lived excited state. After a period of seconds, or even minutes or hours, the quantum is re-emitted as light, this time of a much longer wavelength. This phenomenon, known as phosphorescence, probably occurs only rarely in biological materials.

ANALYTICAL INSTRUMENTS   The simplest kind of colorimetric instrument uses the human eyes as a light detector. The technique consists of comparing the color of the "unknown" solution with the color of a set of solutions of known concentration. The human eye is an amazingly sensitive instrument, and quite small differences in concentration can be detected. In the ordinary practice the "unknown" is held between two of the "known" solutions, and by trial and error we find that pair of standard solutions between which the unknown fits. Various devices have been developed to facilitate this comparison, to reduce stray light, and otherwise to increase the accuracy of the measurement. For the qualitative identification of materials, by locating the black absorption bands it is possible to use a direct vision spectroscope. This is an art which has not been properly developed in the United States but is much more popular among the European workers, particularly several at the University of Cambridge in England.

In the United States we are much more likely to depend upon instruments which contain photoelectric cells to measure light intensities. In order to determine the concentration of a colored material, it is necessary to measure the amount of light actually absorbed by that material. The greatest response is obtained if measurements are made with light of a color strongly absorbed by the molecules. The typical colorimetric instruments isolate a band of wavelengths in the vicinity of this maximum absorption, by means of a system of colored glass filters or by the so-called interference filters. For routine measurements these filter systems are extremely convenient and rapid to use. One merely places the pure solvent in the light path and measures the amount of light striking the photocell. This measurement is often used to establish an electrical zero point. The solvent is then replaced by the colored solution and the diminution in light intensity is noted. The concentration of the solution and the reduction in the light transmitted through it are related, as will be described later. Such an instrument would be called a colorimeter.

SPECTRO-
PHOTOMETERS   Other instruments known as spectrophotometers depend upon a monochromator which uses a prism or a diffraction grating to resolve a beam of white light into its spectrum. The spectrum is allowed to fall upon an adjustable slit which allows light of only one color to pass through. A simplified system is diagrammed in Fig. 9-1. Theoretically it makes no

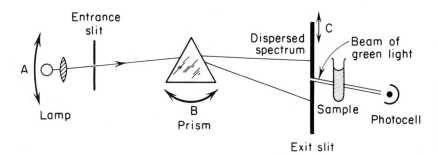

FIGURE 9-1   *A simplified spectrophotometer, set to allow a beam of green light to pass through the exit slit. The wavelength is adjustable in any of the three ways: by rotating the light source through the arc at A, by rotating the prism (B), and by sliding the slit and detector apparatus along the plane at C. The output of the photocell is measured by an appropriate electrical circuit.*

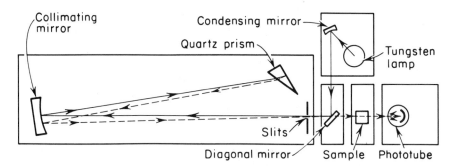

FIGURE 9-2   *Diagram of Beckman Model DU Optical system. Light from the*
tungsten lamp is focused by the condensing mirror and directed
in a beam to the diagonal slit entrance mirror. The entrance
mirror deflects the light through the entrance slit and into the
monochromator to the collimating mirror. Light falling on the
collimating mirror is rendered parallel and reflected to the quartz
prism where it undergoes refraction. The back surface of the
prism is aluminized so that light refracted at the first surface is
reflected back through the prism, undergoing further refraction
as it emerges from the prism. The desired wavelength of light is
selected by rotating the Wavelength Selector which adjusts the
position of the prism. The spectrum is directed back to the col-
limating mirror which centers the chosen wavelength on the exit
slit and sample. Light passing through the sample strikes the
phototube, causing a current gain. The current gain is amplified
and registered on the null meter. (Courtesy Beckman Instrument
Co.)

difference whether the colored solution is placed in front of
or behind the prism. In practice it is usually placed as shown
in the diagram. For comparison and to show some of the
operating details, the optical system of the Beckman Model
DU Spectrophotometer is shown in Fig. 9-2. It will be seen
that the principle is the same. Many models of spectrophotometers are available on the commercial market. They
differ chiefly in the quality of the optical system and in the
electrical measuring system. The smaller and less expensive
instruments typically use a direct reading galvanometer of
some sort. The deflection of this meter is proportional to the
output of the photocell. The more refined instruments use
a Wheatstone bridge as a null measuring instrument, as
described in Chapter 13.

The design and construction of spectrophotometers is an extremely complex subject, so we can discuss only the most gross aspects. The useful range, accuracy, precision, and convenience of a spectrophotometer depend upon the source of radiant energy, the optical system for focusing beams and for dispersion of the spectrum, the light-detecting devices, and the electrical system.

ENERGY SOURCE   The source of radiant energy must produce a continuous spectrum of illumination over the entire range of wavelengths to be used. Ideally, the brightness of the source should be the same, regardless of color. Since this ideal is impossible to achieve, the rest of the spectrophotometer must compensate for differences in the output of the energy source. The lamp must be bright enough to allow the isolation of a narrow band of one color with enough energy to actuate the light-detecting device even after all the reduction in intensity caused by the various elements in the light path. The lamp must be operated from a power supply which does not vary appreciably, so that the energy output of the lamp does not change in time. Usually a tungsten filament lamp is used for the visible region of the spectrum. The lamp is powered by batteries or by electronically-regulated power supplies. A tungsten lamp operating at a different temperature, or an electrically-heated block of metal, may be used as a source of infrared radiation. A gas discharge lamp, usually charged with hydrogen, is used for the ultraviolet region of the spectrum. Hydrogen, when excited by an electric arc, emits a fairly broad, continuous spectrum of ultraviolet radiation. The hydrogen lamp must be encased in a material other than glass since ordinary glass does not transmit much ultraviolet radiation.

MONOCHROMATOR   The optical system focuses a beam of radiant energy from the source, disperses the spectrum of this energy, and focuses a monochromatic beam through the sample to the light-detecting device. A system of lenses can be used for focusing if the reduction in light intensity by the lenses themselves is not objectionable. If the losses are too great, a system of concave mirrors can be used to focus light beams. The spectrum can be dispersed by a prism or by a diffraction grating, each of which has some advantages and some disadvantages. The prism is efficient in that a high proportion of the energy entering the prism is recovered in the single dispersed spectrum. Only one spectrum is produced, but the dispersion is nonlinear; that is, the short wavelengths are separated from each other more than the long wavelengths. The prism is slightly sensitive to temperature, but this is usually unimportant. The diffraction grating produces spectra by the interference principle (Chapter 8) and may operate by either transmission or reflection. Several spectra appear, designated first order, second order, etc. The short wavelength end of the second order spectrum may overlap the long wavelength end of the first order spectrum, necessitating special precautions to eliminate stray light of undesirable colors. The grating, however, is advantageous in producing linear dispersion, which simplifies the mechanical system used to control the wavelength setting of the instrument. Some commercial instruments use prisms; others employ gratings; the Cary Model 14 has one of each.

A set of slits is an integral part of the optical system. A first slit passes a beam of light from the lamp in the prism or grating and blocks out light moving in other directions. A second slit isolates a band of light from the dispersed spectrum. These slits may be adjustable in width.

If the instrument is to be used only in the visible region, glass optical parts are satisfactory. If the instrument is to be useful in the ultraviolet, any elements (lenses, prism) which transmit energy are made of quartz.

RADIATION
DETECTORS   The devices used to detect radiant energy are of several types. Any of these responds more strongly in some regions of the spectrum than in others. Commercial instruments commonly employ two or more detectors of different spectral sensitivity so that the entire spectral range can be covered. The light-sensing device must be able to produce an electrical signal adequate to drive the electrical measuring system. Relatively insensitive detecting and measuring systems are easy to construct and use and are adequate for some purposes. The precision of measurement is limited, since the system will not be able to measure small amounts of energy or small differences in energy. A more sensitive, more precise combination may be desirable, but is usually accompanied by

difficulties in operation, noise, delicacy of control, and more elaborate, expensive circuitry.

Any spectrophotometer must be a properly balanced collection of components. Obviously there is no point in coupling a very sensitive detector to a crude optical system. As a rule, the better the instrument, the narrower the spectral width of the band of monochromatic light that may be used. Imagine a spectrophotometer which produces a spectrum whose energy distribution is shown by the curve in Fig. 9-3. The actual response of the photocell will depend upon the fraction of the area under the curve which passes through the exit slit. For example, the shaded portion between λ_1 and λ_4 would produce a certain electrical response. If the slit is made narrower, a band extending from λ_2 to λ_3 will pass through, and the electrical output is reduced by approximately one-half. As the slit becomes narrower and narrower, the amount of energy falling on the photocell becomes smaller and smaller until eventually there is insufficient energy to produce a usable electrical output. The width of the usable spectral band can be reduced by increasing the responsiveness of the detecting system. The minimum usable spectral band width varies from one part of the spectrum to another, and there is considerable variation among the different commercial instruments. Usually the lower the price, the wider the band widths must be. The effect of spectral band width upon actual measurement is illustrated in Fig. 9-9 on page 135.

RECORDING SPECTRO- PHOTOMETERS

Perhaps the ultimate in convenience is attained in the recording spectrophotometers. These instruments contain a motor-driven mechanism that will scan the entire spectrum, that is, the entire spectrum is swept across the exit slit. At the same time the amount of light falling on the photocell, or some function thereof, is recorded by a moving pen on paper. Several instruments are available on the market, and each uses slightly different principles in its operation. Figures 9-4 and 9-5 compare two of the available instruments. Each uses a double monochromator, or, the beam of monochromatic light produced by one monochromator is passed through a second monochromator to bring about a further reduction in stray light. The instrument in Figure 9-4 operates in the visible and ultraviolet and employs two

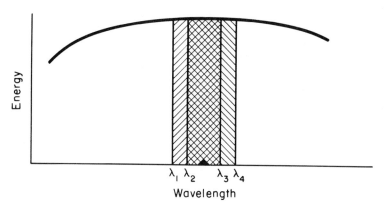

FIGURE 9-3   *Effect of slit width and of spectral band width upon energy passing through exit slit.*

similar quartz prisms. Radiant energy from either the tungsten lamp or the hydrogen lamp enters the monochromator where it is dispersed, and the narrow beam emerges through the exit slit. In the Cary Model 14 radiant energy from either the tungsten source or the hydrogen lamp follows a somewhat similar path, except that the second monochromator employs a grating instead of a prism. The Cary instrument is also equipped to operate in the near infrared. Because of the possibility of producing stray infrared in the measuring system, the infrared beam is produced by a second tungsten bulb and passes backward through the double monochromator. Any stray energy is thus removed.

In any spectrophotometer it is necessary to standardize the instrument at every wavelength with the pure solvent in the light path because the response of the instrument varies with wavelengths. In the recording instruments, the comparison of the sample with the reference solvent must be accomplished automatically since the entire spectrum is swept across the exit slit and the wavelength changes rapidly. Reference to Figs. 9-4 and 9-5 will show two different means of performing this operation. In the first instrument the light beam emerging from the monochromator is chopped by a rotating shutter. This shutter provides pulses of light to the phototubes, yielding a pulsating output which can be amplified. The beam of light is then divided so that half the

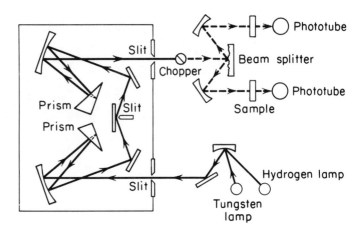

FIGURE 9-4   Optical Schematic Diagram of a Perkin-Elmer recording spectrophotometer. Note the double monochromator and the use of
two matched phototubes as detectors. (Courtesy Perkin-Elmer
Corporation.)

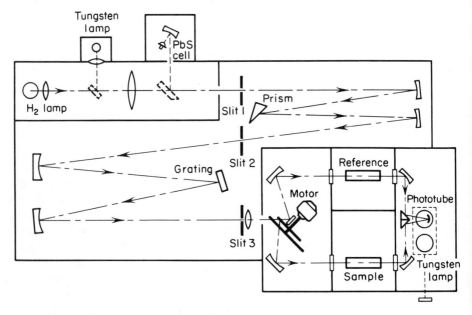

FIGURE 9-5   Optical Schematic Diagram of Cary Model 14 Recording Spectrophotometer, for comparison with Fig. 9-4. Note that the split
beam, after passing through the reference and sample, is recombined in a single phototube. (Courtesy Cary Instruments.)

energy passes through the sample and half through the reference tube, each half-beam arriving ultimately at a separate detector. The outputs from the two detectors are compared electronically, and the difference between the two is plotted on a chart. This arrangement requires a reliable beam-splitter and a pair of carefully matched phototubes but allows the phototubes to be placed close to the sample and reference containers. In the Cary spectrophotometer, a motor-driven shutter-beam-splitter combination provides a time sequence of "dark interval-through reference-dark interval-through sample-dark interval-through reference, etc." Later these two beams are recombined and directed to the same phototube. The two beams are compared by means of an electronic system synchronized with the rotating mirror which splits the beams. This arrangement avoids the possible mismatch of two phototubes but moves the detector farther away from the sample and reference containers.

ULTRAVIOLET AND INFRARED SPECTRO-PHOTOMETERS Measurements in the ultraviolet region require a special source of energy and an optical system which will transmit ultraviolet. Usually quartz optics are used, and the cuvettes or tubes which hold the solutions are made of quartz or fused silica. Otherwise operation in the ultraviolet is the same in principle as operation in the visible region. Ultraviolet spectrophotometers are important in biology because a number of important colorless compounds (proteins, nucleic acids, vitamins, hormones) absorb strongly in the ultraviolet.

Infrared spectrophotometry requires special detectors—often lead sulfide—but an otherwise similar optical system. Absorption in the infrared depends upon different principles than absorption of visible and ultraviolet radiation. Infrared quanta are generally too small to bring about transitions in electronic energy levels. Instead, the quantum size corresponds to the energy differences of the various vibrational states of certain chemical bonds. Most important of these are the heteroatomic bonds, such as $O-H$, $C-H$, and $N-H$. Any molecule which possesses such bonds will absorb infrared radiation at a series of wavelengths. Since the actual collection of absorption bands depends upon the number and kinds of bonds, the infrared spectrum is a characteristic property of a certain molecular structure.

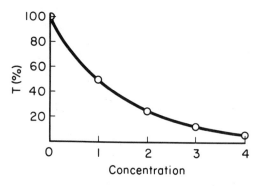

FIGURE 9-6   *Effect of concentration of a colored solution upon the trans-mittancy.*

MEASUREMENTS
OF
CONCENTRATION

We can observe visually that the more concentrated a solution of a colored material, the more light is absorbed and the more intense its color. More formally, however, it is possible to express these relationships by means of equations. Imagine a spectrophotometer which is emitting a beam of mono-chromatic light. We place a container of the solvent in the light path and measure the output of the photocell, or use this amount of light to set the instrument at "100 per cent light transmission." In effect, we are setting the instrument so that it appears that all of the light passes through the pure solvent. With the colored solution in the light path we find that only a fraction of the light strikes the photocell, and the response is accordingly lower. If we refer to the light passing through the solvent as the initial incident light, I_o, and the light passing through the colored solution as I_s we arrive at the following relationship:

$$\frac{I_s}{I_o} = T \qquad (9\text{-}2)$$

where T is the fraction of light transmitted through the colored solution. The value T arrived at in this way (expressed as a decimal fraction or as a percentage) is commonly called transmittancy. It is related to the concentration of the solution according to an equation of the general form $T = k/\text{conc}$. A series of measurements is shown in Fig. 9-6. It would be possible to use this curve in measuring concentrations of unknown solutions. We construct the curve for any particular material at a given wavelength and then, by measuring the transmittancy of an unknown solution and referring to the curve, compute its concentration. This is difficult, however, because of the very curvature of this relationship. A relatively simple transformation of the equation leads to a much more convenient curve, shown in Fig. 9-7. This transformation is shown in the following equation:

$$A = (-\log_{10}\frac{I_s}{I_o}) = \log_{10}\frac{I_o}{I_s} = \log_{10}\frac{1}{T} \qquad (9\text{-}3)$$

The value A has been given various names. It is directly related to the amount of light absorbed and therefore is commonly called absorbancy. It is also a measure of the extinction of the light and is sometimes called extinction. It may also be referred to as optical density. All these terms are in relatively common use but mean essentially the same thing. Determining concentrations in this manner is much more convenient because absorbancy is directly proportional to concentration.

$$A = abc \qquad (9\text{-}4)$$

where a is a constant, b is the length of the light path through the solution, and c is the concentration of the solution. We make a measurement of the light absorbed by a set of standard solutions and then determine the straight line shown in Fig. 9-7.

The amount of light absorbed by a given molecule at any wavelength is a function of the molecular structure. If we measure the amount of light absorbed at any particular wavelength, this can be related to concentration by any of several absorption coefficients (a, above). One of these, the specific absorption coefficient (α), tells us the amount of light absorption to be expected from a solution whose concentration is measured in grams per liter. The molar extinction coefficient (ϵ), on the other hand, tells us the amount of light absorption to be expected from a solution whose concentration is expressed in moles per liter. The value b in the equation must be included because the thicker the solution the more light it will absorb. In many of the commonly used instruments the glass tubes or cuvettes have a light path of one centimeter, and b simply drops out of

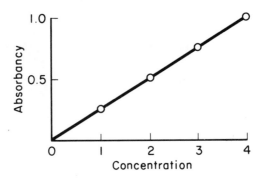

FIGURE 9-7   *Relationship between absorbancy and concentration.*

the equation—but should not be forgotten. Further descriptions of the use of these measurements are given in the following section.

USE OF THE   The spectrophotometer is commonly used for the qualitative
INSTRUMENTS   identification of materials. Since the amount of light absorbed
depends upon the electronic displacements possible within
the molecule, the various colors of light absorbed are a
strict function of molecular structure. Many metals or their
vapors produce sharp-line spectra, but most colored organic
compounds are complicated in structure, and the absorption
bands are spread out laterally. If we wished to know whether
two similar-appearing solutions were actually identical, we
could measure the absorption of different colors of light by
each. The absorption spectrum is a measurement of the
absorption by the colored solution over a range of wavelengths. Wherever the electronic transitions are possible the
molecule will absorb light, and where they are impossible
relatively little will be absorbed. The actual positions of the
maxima and minima depend on the type of molecule. Figure
9-8 shows curves for two pink pigments extracted from
different species of algae. Both solutions were pink although
there was a noticeable difference in the shade of pink.
These two pigments must be similar in structure because
two of the peaks correspond. The third peak in one is
completely missing from the other, however, so the molecules
cannot be identical. Absorption spectra for many kinds of

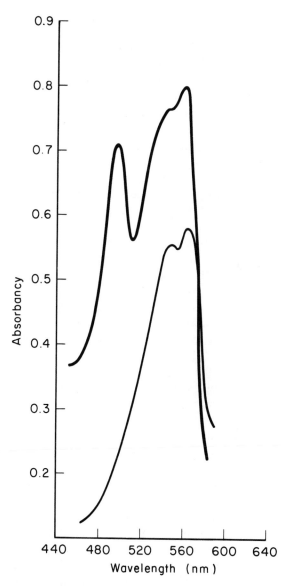

FIGURE 9-8   *Absorption spectra of pink pigments extracted from two different species of algae, traced from a record measured by the Cary Model 14 recording spectrophotometer.*

materials appear in the literature. It is possible to identify materials by comparison with these published curves.

The quality of the spectrophotometer affects the absorption spectrum of a solution in a way that is surprising at first. Narrow spectral band widths produce sharper spectra, as can be seen from Fig. 9-9. One curve was determined by using quite broad bands of monochromatic light. Although the instrument may seem to be set at a certain wavelength, what the photocell sees is really that wavelength plus a band on either side. Thus the apparent absorption at this wavelength is really the average absorption over a wider band. If the actual curve happens to be passing through a maximum at this point, the measured absorption (average) is lower than it should be. The maximum absorption seems to be reduced, and minima are increased.

In measurement of concentrations of materials it is usual to set the instrument at a point of maximum absorption. At this wavelength the reduction in light intensity caused by the absorption by the pigment is considerably higher than the reduction in light intensity from other possible causes, such as scattering. In addition, at these maxima a difference of one or two nanometers in the setting of the wavelength on the spectrophotometer makes relatively little difference in the amount of light absorbed. If, alternatively, the wavelengths were set on a steep portion of the absorption curve, then one or two nanometers could make considerable difference in the absorption. For routine measurements of concentration it is common practice to make up a standard curve from freshly prepared solutions of known concentrations. A curve such as that shown in Fig. 9-7 is drawn, from which the concentration of unknown solutions can be determined. If suitable care is taken in the measurements of the standard solution, it is possible to calculate the specific or molar absorption coefficient from the slope of the curve. If these values are known with precision, then it is not necessary to draw the curve. One merely measures the light absorption, uses the proper absorption coefficient, and calculates the concentration of the solution.

One of the assumptions involved in the development of spectrophotometry was that the amount of light absorbed by a colored material would be independent of the presence of other materials. Each molecule would behave as an individual and would not be influenced by other molecules. Because of the apparent validity of this assumption, it is sometimes possible to measure the concentrations of two

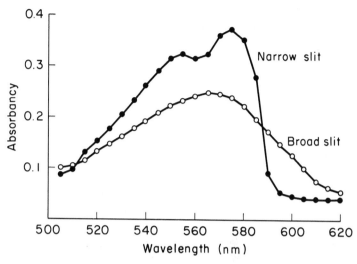

FIGURE 9-9   *Absorption spectra of extract like that shown in Fig. 9-8, measured on two different instruments differing principally in effective spectral band width. (Less concentrated solution than in Fig. 9-8); student data.*

components in the solution by making measurements at two wavelengths. As an example, consider the pair of hypothetical solutions in Fig. 9-10. Pigment P absorbs strongly at 600 nm, while pigment Q absorbs strongly at 650 nm. The absorption by pigment P is low at 650 nm, as is that of pigment Q at 600 nm. The dotted curve in this graph shows the absorption spectrum that would be found in the case of a mixed solution containing equal concentrations of these two pigments. At wavelength 600 nm, a large portion of the absorption results from pigment P. Pigment Q contributes only a small amount to the total absorption. At 650 nm, the relationships are reversed. If we know the absorption coefficients, it is possible to calculate the concentration of both components in this mixture by the following set of transformations. Once the calculations have been made the final pair of equations can be used directly in analyses.

A_{600} = Measured absorbancy at 600 nm.

K^{P}_{600} = Absorption coefficient for pigment P at 600 nm.

$[P]$ = Concentration of pigment P.

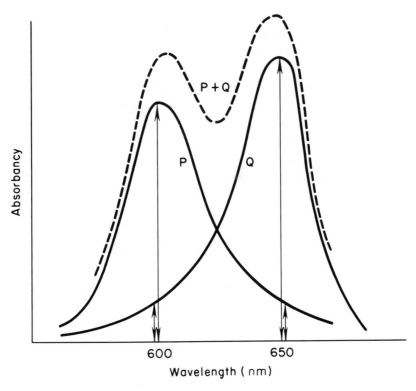

FIGURE 9-10   *Absorption spectra of a pair of hypothetical pigments, and the spectrum to be expected from a solution containing equal concentrations of the two pigments.*

Other symbols follow the same system.

$$A_{600} = K^P{}_{600}\ b\ [P] + K^Q{}_{600}\ b\ [Q] \tag{9-5}$$

$$A_{650} = K^P{}_{650}\ b\ [P] + K^Q{}_{650}\ b\ [Q] \tag{9-6}$$

Solving for [P], from equation 9-5,

$$[P] = \frac{A_{600} - K^Q{}_{600}\ b\ [Q]}{K^P{}_{600}\ b} \tag{9-7}$$

Substituting in equation 9-6 and solving for [Q],

$$[Q] = \frac{1}{b}\left(\frac{K^P{}_{600}\ A_{650} - K^P{}_{650}\ A_{600}}{K^P{}_{600}\ K^Q{}_{650} - K^P{}_{650}\ K^Q{}_{600}}\right) \tag{9-8}$$

Substituting [Q] in equation 9-7 and simplifying,

$$[P] = \frac{1}{b}\left(\frac{K^Q{}_{650}\ A_{600} - K^Q{}_{600}\ A_{650}}{K^P{}_{600}\ K^Q{}_{650} - K^P{}_{650}\ K^Q{}_{600}}\right) \tag{9-9}$$

When the numerical values for the various constants are known, these equations can be simplified.

Another common use of the spectrophotometer is in the determination of reaction rates. In this case we allow the chemical reaction to proceed directly in the cuvettes. In the reaction

$$A + B \longrightarrow C + D$$

suppose component C has a maximum absorption at 365 nm and none of the other components in the mixture, A, B, or D, has any appreciable absorption at this wavelength. If the starting materials A and B are placed in the cuvette in the spectrophotometer and the wavelength is set at 365 nm, the measured increase in optical density will be a measure of the appearance of material C. By carefully timing the reaction we have an adequate and convenient measure of the rate of the reaction.

FLUORESCENCE MEASUREMENTS

Fluoresence is the ability of certain molecules to re-emit absorbed light. Since the molecule remains in the excited state for a short time, the emitted light is of a longer wavelength than that absorbed. With many kinds of compounds, especially at low concentrations, the fraction of the absorbed quanta which appear as fluorescence is constant, or nearly so. This means that solutions of greater concentration will absorb more light and therefore will fluoresce more intensively. In certain situations it is advantageous to measure the fluorescence rather than the absorption. For example, riboflavin absorbs light at the blue end of the spectrum and shows a fairly strong yellow-orange fluorescence. The concentration of a solution could be determined by measuring either absorption or fluorescence. It happens, however, that several nonfluorescent materials commonly present in solutions of this vitamin also absorb appreciably at the same wavelengths. In such a mixed solution it is difficult to ascribe any particular portion of the total absorption to the riboflavin. Since it is the only component with marked fluorescence, measurement of the fluorescence gives a more convenient measure of the concentration.

Several fluorimeters are available commercially. Generally the light beams are handled according to some modification of the diagram in Fig. 9-11. The exciting beam enters the container from one side. The fluorescent light passes through

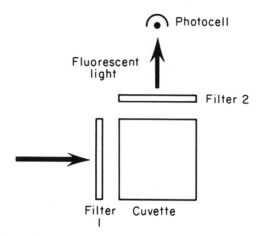

Photocell

Fluorescent
light

Filter 2

Filter

Cuvette

FIGURE 9-11 *A simplified diagram of a system for measuring the fluorescence of solutions.*

a filter, which cuts out the exciting light. It then is measured by a photocell placed at right angles to the exciting beam. Several of the commercially available instruments also have photocells behind the cuvette so that they can measure the light absorption by the sample as well as the fluoresence.

Fluorescence spectra are characteristic of the fluorescing molecule. The instrument used for measuring a fluorescence spectrum must, of course, be more complicated than that for measuring concentration. The fluorescent light is collected and resolved by a monochromator. The intensity of this light at various wavelengths is measured photoelectrically.

FLAME PHOTOMETRY The atoms or ions of metals characteristically absorb and emit light with sharp-line spectra. The flame photometer is used for determining the concentrations of such metals in solutions. The solution to be analyzed is sprayed into a gas flame where the metal atoms are heated until they glow in their characteristic color. Sodium ions, for example, change the colorless gas flame to a brilliant yellow-orange by emitting a strong double line at about 589 nm. The light from the flame is collected and passed through a filter system or monochromator, which passes a band of wavelengths including the bright emission lines of the metal being analyzed. Other wavelengths are removed by the mono-chromator or filter so that only light emitted by this particular metallic atom is allowed to fall on the light-detecting device. Under otherwise identical conditions, a more concentrated solution of the metallic ions produces a brighter light so that measurement of the light passing through the monochromator indicates the concentration of the solution. Instruments are available which are used exclusively as flame photometers, or a gas-burner attachment is used as a substitute for the usual energy source in one of the regular spectrophotometers. Flame photometry is specific; that is, it analyzes one metal even in the presence of others. It is also able to detect very small quantities of a metallic ion (quantities in the general range of milligrams per liter).

PROBLEMS   1. Select a dye of known formula weight, make up a set of solutions of known concentration, and determine the absorption coefficient at the wavelength of maximum absorption. If more than one spectrophotometer is available, especially if the instruments differ in effective spectral band width, compare the values obtained on the different instruments.

2. Living things contain many kinds of pigments. Isolate and purify such a pigment and determine its absorption spectrum, absorption coefficient, and any other spectral properties that seem pertinent.

3. One of the assumptions of spectrophotometric theory is that light absorption by a given colored material is not influenced by the presence of another kind of colored material. Design and carry out an experiment to test this assumption.

SELECTED REFERENCES   Clark, George L., ed., *Encyclopedia of Spectroscopy*. New York: Reinhold Publishing Corporation, 1961. A complete reference work with articles written by competent authorities.

From the Bibliography:

Newman, David W., ed., Instrumental Methods of Experimental Biology.

Strobel, Howard A., *Chemical Instrumentation*.

Willard, H. H., L. L. Merritt, Jr., and John A. Dean, *Instrumental Methods of Analysis*. Almost one-third of this book is devoted to various optical and photometric methods of analysis and the instruments in common use.

Frequently it is desirable to follow the progress of some biological reaction in which gases are produced or used. The best-known of these reactions, of course, are respiration and photosynthesis. Carbon dioxide and oxygen are exchanged in both of these processes.

Respiration occurs in all living cells and converts chemical energy to a form usable by the cell. Some form of food, commonly the carbohydrate glucose, is degraded to carbon dioxide which contains less potential energy than the carbohydrate. In a sense, the production of carbon dioxide is incidental to the more important energy release. When oxygen is available, most cells use oxygen and oxidize the sugar completely to carbon dioxide and water. We might use the imaginary but useful expression $\{CH_2O\}$ to represent one carbon-atom's-worth of carbohydrate. Then a summary equation for respiration is as follows:

$$\{CH_2O\} + O_2 \longrightarrow CO_2 + H_2O$$

If we wish to follow the progress of this reaction we could measure any of the materials in the equation, at least in theory. However, measuring the disappearance of carbohydrate is difficult, largely because the common techniques interfere with the respiratory reaction itself. Measuring the production of water is also unsatisfactory because the cell contains a very large amount of water. The water produced in respiration represents a small change in a large amount of water already present. Generally, the most satisfactory method is to measure the carbon dioxide or oxygen or both of these gases.

The same general principles apply to photosynthesis. Green cells, when exposed to light, perform this reaction which is the reverse of respiration as far as net results are concerned,

$$CO_2 + H_2O \longrightarrow \{CH_2O\} + O_2$$

In photosynthesis the very same gases are exchanged and either of these, or both, can be measured to trace the progress of the reaction.

Respiration and photosynthesis as described here represent only summaries of complicated sequences of single reactions. Most of the individual reactions are enzyme-controlled. Many of the enzymes can be isolated from the cell and can catalyze the same individual reaction under artificial conditions. If one of these separate reactions involved the production or use of carbon dioxide or oxygen, there is no reason that we could not use the same method of measurement used for the whole process.

Gases like carbon dioxide and oxygen can be measured in a variety of ways. Some methods may depend entirely upon chemical principles, others upon physical principles. As an example of a chemical procedure, the carbon dioxide produced by cells may be analyzed by absorbing it in a solution of alkali to produce carbonate,

$$CO_2 + 2\ KOH \longrightarrow K_2CO_3 + H_2O$$

By titrating with a standard acid we can determine the amount of alkali neutralized, and from this amount we can calculate the amount of carbon dioxide. A slight modification of this method depends on the fact that divalent bases like calcium and barium form insoluble carbonate precipitates. From the weight of the precipitate, the amount of carbon dioxide can be calculated. However, these methods frequently are cumbersome and may not be easily adaptable to continuous measurements of gas exchange. Sometimes the chemical techniques destroy the gas being measured, and this destruction might be undesirable in certain types of experiments. There also are several chemical methods for analyzing oxygen, but these are even less convenient than the carbon dioxide analyses.

THE MANOMETRIC TECHNIQUE

The manometric method of measuring rates of metabolic gas exchange is used in almost every cell physiology laboratory in the world. Instruments similar in principle were developed by Barcroft and Haldane about 1902, but Otto Warburg, the noted German biochemist, demonstrated the general applicability of these principles to respiration and photosynthesis. He was largely responsible for the wide adoption of the manometric technique, and the method now bears his name. One should not be surprised to hear, "How did you measure oxygen?" "Warburg." Although we might expect more detailed communication between scientists, this brief answer has conveyed sufficient information.

The principle of manometry is relatively simple. We merely follow the increase in pressure in a closed container as a gas is produced. The behavior of the gas obeys the physicist's gas law, which is conventionally expressed by the following equation:

$$PV = nRT. \tag{10-1}$$

Here P is the pressure, V is the volume, n is the number of moles of gas, and T is the absolute temperature ($^\circ K$). R is the "gas constant," which specifies the relationships of the other items.

If the amount of gas (n) does not change, the equation becomes equivalent to Charles' or Gay-Lussac's law,

$$\frac{P_1 V_1}{P_2 V_2} = \frac{T_1}{T_2} \tag{10-2}$$

It tells us that if the amount of a gas and the temperature remain constant, then an increase in pressure must be accompanied by a decrease in volume. Or if the temperature and volume remain constant, a decrease in the amount of gas will be accompanied by a decrease in pressure. This latter case is used in the manometric method. Temperature and volume are held constant, and the pressure is allowed to vary as the gas is produced or consumed.

In practice, we place the living cells or other experimental material in a small glass container or vessel that can be coupled to a glass U-tube, the manometer (see Fig. 10-1). A change in pressure will be indicated by a difference in the height of fluid in the two arms of the U-tube. The glass tubing is graduated in millimeters so that we can read the height of the fluid.

A glass stopcock in the manometer allows us to leave the vessel open to the atmosphere until we are ready to start the measurements—which means that the pressures inside and outside the container will be equal. When we close the stopcock the living cells will be in a closed space on one side of the manometer. The other side of the manometer is open to the air. Thus we compare the pressure inside the vessel with that outside. As the experiment progresses, the pressure inside the vessel becomes greater than, or less than, that outside.

A reservoir of manometer fluid in a short piece of rubber tubing at the bottom of the U-tube permits the adjustment of the fluid level and allows us to bring the internal volume back to its original level. If we allow the volume to increase, or decrease, it becomes very difficult to calculate just how

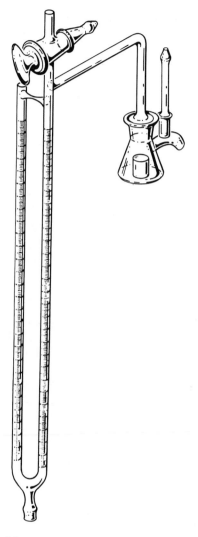

FIGURE 10-1   *Manometer and vessel. A flexible reservoir of fluid is attached at the bottom of the U-tube.*

much gas has been produced or used. Imagine the complexity of calculation in the equation above if P, V, and n all change rapidly. At each reading, the fluid level is returned to the starting point on the closed arm, and the pressure difference is read from the open arm of the U-tube.

The manometer and vessel are conveniently mounted on

a supporting rack. The temperature in the vessel is maintained at a constant value by immersing the vessel in a thermostatically controlled water bath. The water bath is usually the most expensive part of the entire apparatus. A sensitive mercury "thermoregulator" operates through relays to turn on heaters if the temperature falls. Often a refrigeration system is included. The cooling system of course is essential for experiments conducted below room temperature. The thermoregulator balances the heaters against the cooling system to keep the temperature constant. A stirring device agitates the water vigorously so that the temperature is uniform throughout the bath. Water baths differ in detail (see Fig. 10-2), but all of them maintain the temperature within a very narrow range, perhaps at $\pm 0.05°$ C. The water bath system also includes a means of holding the manometer racks and a means of shaking the vessels while they are submerged. The shaking mixes the biological material thoroughly and assures the even distribution of gases within the vessel.

About the time we start to make measurements, a complication becomes obvious. The calculations above assume ideal conditions. Remember that one arm of the U-tube is open to the atmosphere so that we can compare the pressure inside the vessel with the pressure at the beginning of the experiment, that is, the atmospheric pressure in the room. What happens if the pressure in the room changes? Must we abandon our experiment if a thunderstorm approaches and the barometer falls? No, we simply use an extra manometer containing no living material. This manometer can be used to correct for any changes in atmospheric pressure or any slight variations in temperature. This thermobarometer, as it is called, is placed on the water bath along with the experimental manometers. If some variation in the conditions causes a change in the thermobarometer, we simply assume that the other manometers would be affected in the same way. For example, if in one ten-minute interval the thermobarometer rises 2 mm, then each of the other manometers would rise 2 mm in the same time, even without living material. Since the thermobarometer is measuring a change in pressure, it is legitimate to correct the pressure changes of the experimental manometers by adding or subtracting the number of millimeters observed on the thermobarometer. Thinking about it for a few seconds tells us in which direction

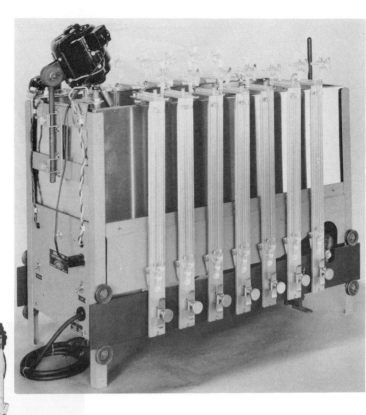

FIGURE 10-2   *Two types of constant temperature water baths. Top, courtesy*
American Instrument Co., Inc. Bottom, courtesy Bronwill
Scientific Division, Will Corporation.

the correction must be applied. An example is given on page 216.

In an actual series of measurements by the manometric method, the results we obtain will be in millimeters of manometer fluid. The measurements would be more useful if they could be expressed as a real amount of a particular gas. "Microliters of oxygen" and "moles of carbon dioxide" have more meaning than "millimeters of manometer fluid." Fortunately, because there is a direct relationship between the pressure change and the change in the amount of gas, we can make the calculations easily. The manometric method is used for small quantities of biological material and measures small quantities of gas. Therefore, the most convenient unit for the gas is the microliter (μl). If we let h represent the number of millimeters of manometer fluid (corrected by the thermobarometer reading) and X the actual amount of gas in microliters, then

$$X = kh \qquad (10\text{-}3)$$

or X is directly proportional to h. It can be seen from the gas equation (10-1) that the relationship will take this form if V and T are constant. If we know the value of the constant k, we can easily find the amount of gas corresponding to any change on the manometer. The value of k will depend upon the conditions under which the experiment is performed and upon the characteristics of the manometer-vessel combination. If the vessel is small, a small change in the amount of gas will make a large change in the pressure and in the reading of the manometer. In contrast, if the vessel is larger, the same amount of gas will make a smaller change in the pressure.

In order to compare the results of different experiments, we express the amount of gas used or produced with reference to a set of standard conditions, $0°$ C ($273°$ K) and atmospheric pressure (760 mm Hg). Chemists very often express amounts of gas in these terms because they know that one mole of any gas, under these conditions, will occupy 22.4 liters, or any given volume of gas under these conditions will contain a specific number of moles. Although it may seem difficult to transform manometer readings to volume of gas under "standard conditions," we can incorporate these corrections during the computation of the constant k. An equation has been developed such that the constant k

transforms h millimeters of manometer fluid directly into microliters (at $0°$ C and 760 mm Hg) of the gas being measured. The complete equation follows:

$$k = \frac{V_g \dfrac{273°}{T} + V_f \alpha}{P_0} \tag{10-4}$$

$V_g =$ volume (in μl) of the gas space in the particular vessel-manometer combination.

$T =$ absolute temperature, or Celsius temperature $+273°$.

$V_f =$ volume (in μl) of liquid in which the living cells are suspended.

$\alpha =$ solubility of the gas in this liquid at this temperature.

$P_0 =$ the number of mm of this manometer fluid that would exert the same pressure as 760 mm of Hg. In other words, this is the "standard pressure" in terms of millimeters of manometer fluid.

For any separate vessel-manometer combination we must measure the internal volume. The easiest accurate means of finding this volume is to find how much liquid the vessel will hold. Usually we fill the gas space with mercury and then weigh the mercury. The density of mercury is known quite precisely. Since it is great, a small difference in volume brings about a large easily-measured change in weight.

The solubility factor, α, may be found in physical tables, or more easily, in the Umbreit, Burris, and Stauffer book, *Manometric Techniques.* It must be included in the equation because a certain amount of the gas we wish to measure will remain dissolved in the liquid.

In the biological experiments, P_0 is usually 10,000 mm of manometer fluid. This round number is no coincidence, because the manometer fluid is a specially prepared solution whose density will be 760/10,000 of the density of mercury. The fluid most commonly used is Brodie's solution, 23 g of sodium chloride in 500 ml of water. A small amount of detergent is added to prevent the liquid's sticking to the glass tube. The solution is colored with a dye for ease in reading.

Suppose that we wish to measure oxygen exchange at $25°$ C in one of our manometers. We have found that the volume of the internal gas space is 18.36 ml (18,360 μl). For convenience we prefer to use 3 ml (3000 μl) of fluid. Thus $V_f = 3000$ μl, and V_g is the total volume minus V_f,

or $18{,}360 - 3000 = 15{,}360$ μl. For oxygen in water at
$25°$ C, α is found to be 0.028. If we use Brodie's solution,
P_0 is 10,000 mm of manometer fluid. The arithmetic is not
difficult. We find that the constant for oxygen *in this vessel
at this temperature* is 1.42 μl of O_2/mm. This means that
each millimeter on the manometer corresponds to 1.42
microliters of oxygen.

If we keep our slide rule handy, it is an easy matter to find
the actual amount of oxygen, even while the experiment is
progressing. If we wished to measure CO_2, we could find
another constant by substituting the proper solubility figure
in the equation. The example given here is a reasonably sized
constant. Most of the vessels in common use today are of
such a size that the constants for oxygen are between 1.00
and 2.00 μl O_2/mm.

The actual measurement is accomplished by placing the
necessary materials in the vessels and placing the vessels on
the manometers. The manometers are placed on the water
bath with the vessels hanging into the water. With the
manometer stopcocks open, a period of 15 to 20 minutes is
allowed for temperature equilibration. Then, as rapidly as
possible, the manometers are closed, and the initial pressure
and starting time are recorded. The number of millimeters of
pressure change is recorded at five or ten minute intervals.
The final treatment of data from manometric measurements
is used as an example in Chapter 15.

In principle, the production or use of any gas could be
measured by this technique. Actually carbon dioxide and
oxygen are the gases most frequently involved in biological
reactions. Unfortunately, in respiration and in photosynthesis
one of these gases is utilized while the other is produced.
There is no net change, and the manometer will show no
pressure change. Obviously a solution for this complication
must be found. Some quite simple remedies are in common
use, as well as some that are fairly complex. Perhaps the
most obvious answer to the problem is to prevent the change
in one of the two gases. If we wish to know the rate of the
whole respiratory reaction, we may be satisfied to measure
only oxygen. We must assume that respiration is the only
cell process that uses or produces any gas, but often this
assumption is safe enough. We can place the living material
in the main part of the vessel and then, in a separate com-
partment, place a small amount of alkali solution. The carbon

dioxide produced in respiration diffuses through the gas space to the alkali where it is absorbed. Thus the only gas exchange that will affect the pressure is the change in oxygen. In many cases the amount of carbon dioxide produced is about the same as the amount of oxygen used.

In photosynthesis another method is necessary because carbon dioxide is needed in the reaction. Ordinarily this measurement is performed by suspending the green cells in a bicarbonate solution. As the cells use carbon dioxide, more of this gas is provided by the bicarbonate. Since the carbon dioxide concentration in the gas phase does not change, we measure only the production of oxygen. This technique is used quite commonly even though the high *pH* of the bicarbonate solution may influence the activity of the cells. It might seem convenient to absorb the oxygen, just as the carbon dioxide was absorbed in respiration. However, oxygen absorbers are expensive, dangerous or difficult to handle, or react quite slowly. If we wish to avoid the alkaline conditions of the bicarbonate solution, we must use one of the more complicated tricks of the manometric technique.

Anyone who wants to use the manometric technique extensively should consult the standard reference book by Umbreit, Burris, and Stauffer, *Manometric Techniques*. It contains complete discussions of many applications of manometry and a number of variations that are possible. It also incorporates information on several other useful laboratory methods.

This description has been restricted to carbon dioxide and oxygen. However, exchange of other gases may result from some of the metabolic reactions of cells. Some algae, for example, can perform a reaction in which molecular hydrogen is involved. Ammonia, nitrogen, methane, hydrogen sulfide, and perhaps other gases might appear or disappear in certain specialized biological processes. One of the great advantages of the manometric method is that it could be used for any of these.

The techniques of manometry are not especially difficult to learn. The glassware is fragile, but easy to handle, and a variety of kinds of vessels have been developed, including some specialized for certain unusual measurements. Figure 10-3 shows some sample types. Only a little practice is required to read the manometer scales, even while the manometers are moving. In a short time it is possible to learn

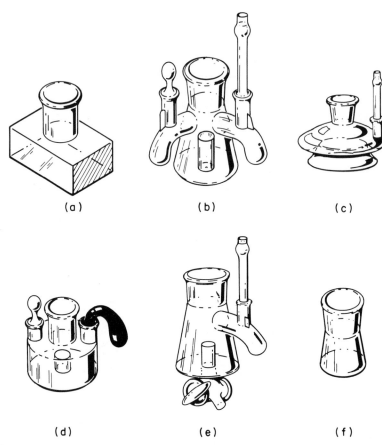

(a)   (b)   (c)

(d)   (e)   (f)

FIGURE 10-3   *Some examples of special manometer vessels. The rectangular*
vessel (a) is often used in photosynthesis studies because a greater
surface can be exposed to light from below. The sidearms, as in
b, permit the addition of materials while the measurement is in
progress. Alkali for trapping CO_2 *might be placed in the center*
well, as in b or d, in a concentric trough, as in c, or added after
the experiment is in progress (e). Number f has a reduced volume
to permit greater sensitivity of measurement.

to recognize the behavior of the manometers when something
is wrong. Occasionally, one of the ground glass joints will
leak, causing pressure changes that appear too low. Sometimes
the temperature of the bath will drift slightly, leading to large
thermobarometer corrections. Rarely, the temperature in

the bath is not uniform from place to place. In this case, the thermobarometer correction may not truly represent the change that would have occurred in the experimental manometers if they had had no living material.

These difficulties are less bothersome than the annoying problems that are associated with many other laboratory instruments. The manometric technique has proved its advantage over many years. The literature contains only a few papers where manometry has been pushed beyond its limitations, a fact which also attests to the value of the method.

Several new physical methods of measuring various gases have been developed recently. Some of these are much more convenient for certain measurements, and some provide a specific analysis for a particular gas. Although they will be used to advantage, it is doubtful that they will ever completely replace the old reliable manometers.

THE INFRARED GAS ANALYZER One of the more popular of the recently developed instruments for gas analysis is the infrared analyzer. Most organic compounds have infrared absorption bands, arising from the vibrations of the $C-O$, $C-H$, $O-H$, and other bonds. Any individual molecule has a definite collection of such heteroatomic bonds and, consequently, a defined spectrum of absorption in the infrared region. This principle is the basis of the infrared spectrophotometry mentioned in Chapter 9.

The Beckman Infrared Gas Analyzer is essentially a colorimeter designed to operate in the infrared region. It can be used for any of several gases, including carbon dioxide, water, and methane. The instrument responds to the amount of a particular gas in a stream of air, providing a continuous record of its concentration.

The essential features and the operating principle can be seen by referring to Fig. 10-4. This is a simplified diagram of the analysis unit. The air or gas to be analyzed is drawn continuously through the sample tube. Infrared from the two sources (S) passes through the sample tube and the reference tube to the detector chambers. The windows (W) transmit infrared but prohibit the free movement of gas. The two detector chambers are filled with the gas to be analyzed, for example, CO_2. As the CO_2 absorbs infrared energy it tends

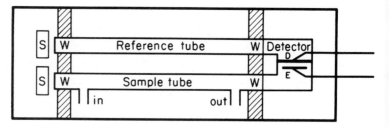

FIGURE 10-4   *Simplified diagram of the detection system in the Beckman Infra-Red Gas Analyzer. (Courtesy Beckman Instruments, Inc.)*

to expand, exerting pressure on the diaphragm (D). With no CO_2 in the sample tube, the instrument is adjusted to indicate equal pressures on the two sides of the diaphragm. Thereafter, any CO_2 in the sample tube will reduce the amount of energy reaching the lower detector chamber, and the resulting pressure differential moves the diaphragm. The diaphragm and the stationary electrode (E) form a condenser whose capacitance changes as the distance between the plates changes. A radio frequency voltage is applied across this capacitor, and a very small movement of the diaphragm produces a usable signal which is amplified and employed to drive a recorder.

The fact that the gas to be analyzed is used in the detector makes the instrument absolutely specific for this gas. Of course, as the reduction of energy in the sample tube depends upon the number of molecules of gas in the light path, the instrument is sensitive to pressure and temperature changes. Temperature control is provided in the instrument, and it is not too difficult to prevent pressure surges in the external gas stream.

This gas analyzer responds extremely rapidly to very small variations in CO_2. The instrument in our laboratories has been used to monitor the air in a room by pumping room air through the sample tube. The analyzer will detect one more person walking into the room in less than a minute.

Analyzer tubes of several different lengths are available to permit the use of the instrument for several different ranges of gas concentration. If the detector chambers are charged with water vapor, we have a useful indicator of humidity. Various industries have similarly used the infrared analyzer to detect methane, ethane, and other gases.

MAGNETIC OXYGEN ANALYZER Another instrument offering a specific test for a certain gas is the magnetic oxygen analyzer. Oxygen is about the only gas which is to any extent paramagnetic; that is, it is attracted into a magnetic field. Carbon dioxide, water vapor, and most other gases likely to be encountered in biological experiments do not respond appreciably to a magnetic field. Several models of oxygen analyzers are available.

The instruments consist of a chamber surrounded by a magnetic field. A stream of the air to be analyzed flows past this chamber, but only oxygen is drawn in. The pressure in the chamber is related to the amount of oxygen in the gas stream. In the Beckman instrument, changes in amounts of oxygen cause a rotation of a small dumbbell suspended on a fine wire. The rotation of the dumbbell is measured directly in some models or indirectly, through a null-point optical and electrical system, in other models. Another instrument, made by Siemens in Germany, determines oxygen concentration by measuring heat conductivity in the chamber. Changes in the amount of oxygen affect the conduction of heat from a wire, which in turn influences the electrical resistance of the wire.

In our experience, the Beckman Magnetic Oxygen Analyzer has proved to be slightly slow to respond because the gas must diffuse into the magnetic chamber. Otherwise, the instrument is very reliable, yielding reproducible measurements of oxygen concentration very conveniently.

ELECTROCHEMICAL METHODS For the determination of the utilization or production of oxygen, several electrochemical instruments are available. These instruments, in general, depend upon an oxidation-reduction reaction in conjunction with an electrode which furnishes or withdraws electrons. The principle is basically the same as for the electrode reactions described in Chapter 13. The measuring circuit records an electrical current, i.e., the rate of flow of electrons. Therefore, the rate of a biological reaction such as respiration or photosynthesis can be recorded directly. Separate measurements of the amount of oxygen present and the time, followed by a calculation, are no longer necessary. In many instances, the direct measurement of the rate of the reaction hastens the work of the laboratory.

PROBLEMS 1. You will recall that the manometers described in this chapter operate by measuring the pressure change while the volume remains constant. In recent years manometers have become available in which volume change is measured by a micrometer device and the manometer fluid levels are used only to adjust the pressure to a constant value. In what way must the "manometer constant" be modified in order to calculate the number of microliters of gas exchange under standard conditions from the measured volume changes?

2. In this chapter the major emphasis has been on systems for working with very small amounts of biological material. If you wished to make measurements on a whole large animal, the complete change of scale and other difficulties would make it impossible to employ the general principles described here. If you were contemplating the use of a manometric method on whole animals, what sorts of difficulties would you anticipate? What kind of system might be developed in order to avoid these difficulties?

SELECTED REFERENCES Dixon, Malcolm, *Manometric Methods as Applied to the Measurement of Cell Respiration and Other Processes,* 3rd ed. New York: Cambridge University Press, 1951. An especially valuable discussion of the theory of manometry.

Umbreit, W. W., R. H. Burris, and J. F. Stauffer, *Manometric Techniques,* 4th ed. Minneapolis: Burgess Publishing Company, 1964. The standard laboratory handbook on the subject.

Chromatography is a laboratory technique for separating mixtures of similar materials from each other. A solution of the mixture is allowed to flow with its solvent over the surface of a finely-divided or porous solid material. The different components in the mixture flow at different rates, eventually becoming separated from each other. Chromatography has become popular in biological work because extracts from living cells contain different materials of similar chemical nature. Often the only way to analyze those mixtures is to separate the components from each other. For example, we might break down a protein into the amino acids of which it was constructed. Since all the various amino acids are similar in chemical and physical properties, it may be difficult to study them by ordinary chemical techniques. Chromatography is the most convenient way of separating these amino acids from each other. As a second example, some of the work on the metabolism of cells involves mixtures containing simple sugars and sugar derivatives. Such mixtures also are separated by chromatography. A third example is a separation of the chloroplast pigments of plants. These materials are highly colored, and it is easy to follow the progress of the separation.

Chromatography means literally "writing with color." The name was chosen because the method was first developed to separate mixtures of colored compounds. Several different physical principles are involved, but a discussion of these principles is better delayed until after a description of the method.

CHROMATO-
GRAPHIC
PROCEDURES:
COLUMN
CHROMATOGRAPHY

In the following paragraphs, each of several chromatographic techniques is described generally.

Chromatographic separations can be performed with a finely powdered adsorbing material in glass tubes of various sizes and types. A very convenient glass tube is fitted at one end with a ground glass joint and a fritted glass disk. The ground joint permits the easy removal of the solid material when the separation is complete. The fritted disk holds the adsorbing material in the tube while it allows solvents to pass through and drip out at the bottom of the tube.

155   The following description will be easier to follow by referring to Fig. 11-1. This set of glass tubes represents a series of stages in a chromatographic separation.

Preparing a column requires a certain knack, and the beginner should not be discouraged if the first few columns do not work well. Several different methods of packing columns have been described, but the following has produced good results in our laboratories. The bottom of the tube is closed off so that liquid will not leak out. Solvent is poured in until the tube is approximately half full. The powdered solid is sifted into the liquid and stirred with a glass rod to make a very thin paste. Alternate additions of powder and solvent can be continued until the tube is almost full of the thin paste. When the stopper is removed from the bottom of the glass tube, solvent will drip out and the powder will settle into a uniform column. Usually the solid settles faster than the solvent escapes, so a layer of clear solvent appears at the top of the column. Occasionally the stirring rod is used to redistribute any uneven settling. After a short time the solid will have settled as much as it will, and the remaining solvent will percolate downward and out the bottom of the tube. Some people find it more convenient to mix the solvent and solid in a beaker to form a thin slurry, and then to pour this slurry into the tube. There the solid will settle into a uniform column. Still others prefer to pack the column dry and then saturate it with solvent.

When most of the solvent has disappeared from the top of the column, the mixture of substances to be separated is added by allowing it to flow slowly down the side of the tube from a pipette. The mixed materials and solvent will percolate into the upper layers of the column.

When most of the mixture has moved into the top of the column, the column can be "developed" as follows. More solvent is added to the tube, a process also accomplished by allowing the liquid to flow slowly down the side of the tube from a pipette. Certain substances from the mixture will move down the column more rapidly than others. Eventually, well separated bands will form.

The different materials may be recovered in either of two ways. They may be allowed to move downward until they drip out the bottom of the tube. Each component can be caught in a separate container. In this case we are "washing" the materials off the adsorbing substance by the addition of solvent, a process commonly known as *elution*. Mechanical

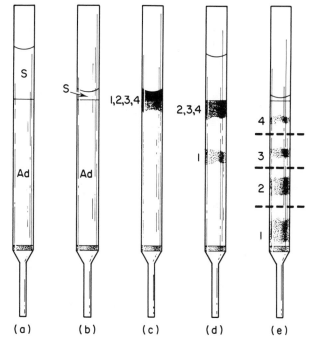

FIGURE 11-1   *A series of chromatograms at different stages of development. In
a, the adsorbent (Ad) has settled, leaving a layer of solvent (S)
on top. A little later (tube b) this layer of solvent has percolated
through the adsorbing material. In tube c, a mixture of four
materials has been added, and has begun to move into the column.
In a few minutes (d), developing solvent has been added, and
solute 1 is moving away from the other materials. The finished
chromatogram (e) shows four separate bands which may be
recovered by slicing the column along the dashed lines.*

fraction collectors which move a series of test tubes under
the column at spaced intervals can reduce the drudgery of
collecting fractions. If the bands of the separated components
can be identified by color, it is usually faster to allow the
materials to become well separated from each other on the
column, and then push the plug of solid from the tube. This
plug can be sliced into the different colored bands. Each slice
is placed in a separate container, and suitable solvent is
added to liberate the separated components from the solid
adsorbing material. We now have several different solutions,
each containing a different component.

Successful chromatograms are possible only on evenly packed columns. The technique of column packing apparently must be learned through several trials.

PAPER
CHROMATOGRAPHY

In paper chromatography, the solid material is ordinary filter paper. Various combinations of solvents can be used to separate different kinds of mixtures.

If we should wish to obtain a quick estimate of just which components were contained in a mixed solution, a paper chromatogram would allow a more rapid determination than would the preparation of a column. The chief disadvantage of the paper technique is that only small amounts of materials can be separated. The following description includes several variations of the use of filter paper for chromatography.

1. A strip of filter paper about 2 cm wide is cut to fit into a large test tube. A hook in a cork stopper suspends the paper in the tube so that the bottom of the paper strip almost touches the bottom of the tube (see Fig. 11-2). A small amount of the solution of mixed components is placed along a pencil line near the bottom of the paper and then dried. A suitable solvent mixture is placed in the bottom of the test tube. The paper and cork stopper are fitted into the tube with the bottom of the paper dipping into the liquid and the spot of mixture about a centimeter above the surface of the liquid. As the liquid rises in the filter paper by capillary action, various components are carried along the paper strip. Some move more rapidly than others, and after a time several distinct bands can be identified.

2. Larger amounts of the mixture can be separated, or several different mixtures can be tested simultaneously, by using a large beaker or battery jar instead of a test tube. A square of filter paper is prepared by placing a line of the mixture near one edge or by placing spots of several mixtures near the edge. The paper is then rolled and fastened into a cylinder with the spots of mixtures at one end. This cylinder will stand upright in the beaker or battery jar (Fig. 11-3). If solvent is placed in the bottom of the glass container, it will rise in the paper just as in the previous example. The same kind of separation occurs.

3. Sometimes the paper chromatogram will develop better if the liquid is allowed to descend over the paper instead of rising by capillarity. A spot of the mixture is placed near the

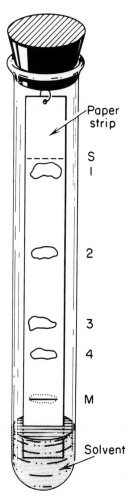

FIGURE 11-2 *Ascending paper chromatography. The mixture of materials was placed along the line (M) and the paper was dried. It was then placed in the test tube, and the solvent ascended by capillary action, having reached point S at the present time. Components 1, 2, 3, and 4 have been carried upward at different rates.*

top of a strip or sheet of paper. This end is placed in a container of the solvent, with the rest of the paper hanging down over the side. As the solvent moves downward on the suspended paper strip, the various components are carried along at different rates. This operation is conducted inside

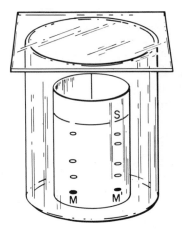

FIGURE 11-3   *A square of filter paper was rolled into a cylinder for this chromatogram. The edge of the paper could be fastened together with staples, adhesive strips, or with thread. Spots of a mixture were placed at the circles marked M, and the solvent has risen to point S.*

a closed container, where the atmosphere is saturated with the solvent vapor.

4. Solvent is placed in the bottom of a round glass dish. A cone of filter paper is prepared and placed point-up in the center of the dish so that the point of the cone stands just higher than the edge of the dish. Now a disk of filter paper slightly larger than the diameter of the dish is prepared by placing a ring of the mixture to be separated around the center of the paper. After drying, the paper is placed over the dish, with the point of the cone piercing the center of the paper disk. An identical glass dish is placed over the paper to serve as a cover. The paper cone acts as a wick to deliver solvent to the center of the paper disk. From this point the solvent moves outward into the paper in all directions. The finished chromatogram will be a circular disk of paper with concentric rings of the various separated components.

5. Paper chromatographic separations can be performed in two dimensions. We place a spot of the mixture of materials at one corner of a square of paper. Then, using one solvent mixture, we allow the various materials to move upward along one edge of the paper. Now we rotate the paper so that the separated spots are at the bottom and run the sepa-

ration again, this time using a different solvent mixture (see Fig. 11-4). The finished two-dimensional chromatogram will have spots distributed in various locations all over the paper because of the different rates with which the individual compounds move in different solvent combinations.

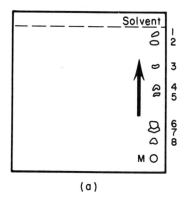

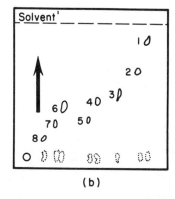

(a)　　　　　　　　　　　(b)

FIGURE 11-4　*Two-dimensional paper chromatography. A spot of the mixture was placed at M, and solvent was allowed to ascend to point S, separating the components of the mixture (a). In step b, the paper was rotated and another solvent was allowed to ascend to level S', bringing about a further separation of the components.*

THIN LAYER CHROMATOGRAPHY　A technique which combines the advantages of column and paper chromatography is the separation on thin layer plates. Originally, and still very commonly, a slurry of a solid adsorbing material, with a binding agent if needed, is spread in a thin film on the surface of a glass plate. In one common procedure, for example, a slurry of specially prepared silica gel, with $CaSO_4$ as a binder, is spread on glass plates 20 cm square. Spreading devices are used to spread the slurry in a film of uniform thickness. The plates are dried, usually in an oven, and then can be handled in much the same way as paper is used in ascending chromatography. After the separation, spots can be recovered by scraping from the glass.

Separation on thin layer plates is accomplished very rapidly. With some mixtures a paper chromatogram may require 24 hours to develop properly. With thin layer chromatography, the same job may be finished in less than an hour. Another advantage is that somewhat larger quantities of the mixture can be separated than on paper.

It is possible to use almost any solid adsorbing material that might be used in a column, with almost any solvent combination that would be appropriate for the mixture to be separated. Several companies now market plastic sheets pre-coated with various adsorbing materials.

Different principles are involved in the different kinds of chromatography. The following physical principles certainly are involved at one time or another, although it may not be easy to tell which is operating in any particular case. Adsorption is a phenomenon which occurs at the surfaces or interfaces between two different kinds of material, as between a liquid and a solid or between a gas and a liquid. A certain amount of energy will be associated with this interface, as can be seen from the unequal energy distribution which causes the surface tension of water in contact with air. Some dissolved materials tend to reduce the amount of interfacial energy, in which case, they collect at the interface. Charcoal is useful for decolorizing solutions, because the colored materials are adsorbed on the charcoal, or for purifying air, because certain gases are adsorbed.

Adsorption certainly is involved in at least some forms of chromatography. For example, in the separation of chloroplast pigments on columns of sucrose the various pigments are bound (adsorbed) on the surfaces of the sugar particles, but different pigments are held with different degrees of tenacity. Each of the pigments has its own char-acteristic solubility in the solvent also. When the mixture of pigments is placed on the top of the column, the molecules are bound by the adsorbing material (sugar). A fraction of the molecules escape from this binding, however, and are carried downward with the solvent, only to be bound again in another location. The binding strength and the solubility in the solvent determine how far the average molecule travels before being bound again. The probability of moving rapidly or slowly is different for each component. If the solvent is modified, these relationships can be changed slightly.

LIQUID-LIQUID PARTITION
Whenever a material is placed in a mixture of two solvents which are insoluble in each other, the dissolved material distributes itself in the two solvents according to a definite

relationship. For example, consider a solute X which will dissolve in solvent A as well as in solvent B. If solvents A and B, immiscible in each other, are placed in a container and subsequently a small amount of X is added, part of X will go into each solvent. An equilibrium will be reached when a certain constant fraction is contained in each solvent. The ratio of the concentrations in the two solvents is called the partition coefficient:

$$K = \frac{\text{conc. in solvent } A}{\text{conc. in solvent } B}$$

Suppose that this ratio is large, meaning that most of the solute goes into solvent A. Given a solution of X in solvent B, the material X could be transferred to solvent A by successive additions of fresh solvent A.

In another instance, imagine that X distributes itself 5 per cent in water and 95 per cent in hexane. The two solvents are immiscible. Another material, Y, reaches its equilibrium when 90 per cent is in water and 10 per cent in hexane. Given an aqueous solution containing both X and Y, the two solutes could be separated from each other by mixing the solution with hexane. At equilibrium, the water would contain only 5 per cent of the X but 90 per cent of the Y. If this water layer is drawn off and mixed with fresh hexane a new equilibrium will be established. The water then will contain only $0.05 \times 0.05 = 0.0025$ of the X and $0.90 \times 0.90 = 0.81$ of the Y. After another step or two, X virtually disappears from the water, but only a little of the Y is lost. If the components in a mixture differ greatly in their partition in the two solvents, this is a practical method of separation.

If the components of a mixture differ only slightly in their concentrations in two immiscible solvents, or if there are several different solutes in the mixture, this method becomes impractical. If one of the solvents is bound on the surfaces of a solid material, however, and the other solvent is allowed to flow over it, even relatively similar solutes may separate from each other. In most of the systems of paper chromatography one of the individual solvents in the mixture is likely to adhere as a film on the paper, while the other solvent flows over this film. The paper merely serves as a base on which the natural distribution or partition can occur. Since in some cases chromatographic separations can occur on dry paper with single solvents, apparently adsorption is also involved.

ION EXCHANGE
CHROMATOGRAPHY

Materials which have the ability to dissociate into positively and negatively charged ions can sometimes be separated by virtue of the electrical charge. A few natural minerals and a large number of synthetic resins bind ions on the surfaces of particles. More recently, several charged carbohydrate derivatives have been used. A resin might consist of a substance containing, for example, a number of acidic groups as part of the molecular structure. When the resin is in water, these acidic groups ionize, leaving negatively charged spots on the resin particle. Cations will be held at these spots, some more tightly than others. If a solution containing a mixture of cations is poured over this resin, these positively charged ions, such as Na^+, K^+, or Ca^{++}, will displace hydrogen ions from the resin. In effect, the resin exchanges its H^+ ions for metallic ions.

If a mixture of positively charged ions is allowed to flow continuously over such a resin in a long column, some ions will be carried along with the water faster than others. Ions which form a strong electrical bond with the negatively charged radicals on the resin will move very slowly. If the column is long enough, the various kinds of ions will emerge at the end of the column one at a time.

A fairly complex set of equilibria exists in an ion exchange column. Imagine a synthetic resin which binds Ca^{++} ions more strongly than Na^+ ions and Na^+ more strongly than H^+ ions. If the resin exists almost entirely in the acid form, that is, holding H^+ ions, and a solution of Na^+ ions is added, the Na^+ ions displace hydrogen ions from the resin. Later, the addition of Ca^{++} will cause the removal of Na^+ and the binding of Ca^{++}. The binding is not permanent, however, and a large excess of hydrogen ions, as from HCl, could cause displacement of the Ca^{++} ions and thus bring about the regeneration of the original resin.

Other resins, themselves positively charged, attract negatively charged ions. A mixture of organic acids might be separated on such a resin.

The theoretical treatment of the elution of materials from ion exchange columns becomes quite complex, and there is no complete agreement on the principles involved. The synthetic resins are extremely useful in preparing ion-free water and in separating mixtures of amino acids, nucleic acids or nucleotides, and carbohydrate derivatives. The tradenames Dowex and Amberlite have become very familiar terms in the modern laboratory.

GEL FILTRATION Gel filtration is a relatively new form of chromatography, employed most commonly in columns, occasionally on thin layer plates. Sephadex is the trade name of a solid adsorbing material which consists of macromolecules of a carbohydrate crosslinked to each other to form a three-dimensional network. Different degrees of crosslinking have been worked into the different grades, thus producing networks with large or small pores. Separation of mixtures occurs by a "molecular sieve" effect. Small molecules can penetrate into the pores of the solid; large molecules are excluded entirely. If a mixture of materials is placed on the column and followed by water or a buffer solution, the largest, excluded molecules will be eluted first. Thereafter, the smaller molecules will be sorted on the basis of size, the smallest ones appearing last. Since molecular size is closely related to molecular weight, gel filtration is often said to sort molecules by molecular weights. Different Sephadex gels are available, covering different molecular weight ranges. The most tightly crosslinked product separates molecules in the molecular weight range from 100 to 5000. Other Sephadex preparations separate proteins with molecular weights up to about 400,000. New agarose preparations separate polymolecular particles with weights up to several million.

PRACTICAL CHROMATOGRAPHY A great variety of mixtures may be separated by chromatography. All that is required is a combination of solvents and solid materials such that there is a difference in the physical properties of the various components of the mixture. Whenever a new or untried mixture is to be chromatographed, the investigator must choose from among paper, adsorbent columns, ion exchange resins, and gel filtration. The filter paper technique, easiest and quickest, is usually tried first. The next question concerns the type of filter paper to use. The manufacturers produce a variety of papers, some of which may work better than others in any particular chromatographic separation. It is convenient to keep samples of various types in the laboratory, and a few preliminary trials usually will show that one paper excels the others in speed or completeness of separation.

The solvent to be used for development must also be chosen carefully. Each mixture of materials has its own set of properties, a fact which has an effect on the choice of solvents. If the general class of compounds to which the mixture belongs is known, the literature or experience will suggest several solvents. The materials to be separated must not be too soluble in the solvent or they will move together almost as rapidly as the solvent moves. If, in contrast, the solubility is too low, they will remain at the spot where they were originally placed.

When colorless mixtures are separated, the spots must be identified somehow. Amino acids, sugars, and many other colorless compounds can be separated on paper, but the individual spots or bands must be found later. Sometimes the paper is treated with a reagent that reacts with the separated materials to yield a colored product. Some materials are fluorescent under ultraviolet illumination, while others absorb ultraviolet and appear as dark spots on a slightly fluorescent paper. If some components in the mixture are labeled with radioactive tracers, the paper chromatogram will "take its own picture" if clamped against a sheet of photographic film according to the technique known as autoradiography.

If the separated compound is to be recovered unchanged from the paper, it may be necessary to run two chromatograms under identical conditions. One is treated with a reagent to find the spot; the unchanged molecule is recovered from the same spot on the other chromatogram.

A numerical value, R_f, is frequently used in locating certain compounds.

$$R_f = \frac{\text{rate of movement of solute}}{\text{rate of movement of solvent}}$$

On paper chromatograms, different components of a mixture move at different speeds. The "front" of each compound moves at a certain fraction of the speed at which the solvent moves. If one amino acid moves very rapidly in a certain solvent mixture, its R_f value will be high, perhaps 0.85. Another amino acid might move very slowly in the same system and have an R_f value of 0.15. In reporting R_f values it is important to specify the conditions of the separation and whether ascending or descending chromatography was used.

On columns and on paper, any moving component commonly moves as a mass of molecules with a "tail" trailing out behind. This tail often will be mixed with the front of the next component. The second component in a mixture is thus harder to purify than the first. Sometimes this difficulty can be overcome by changing solvents and reversing the order of the two compounds.

It is frequently possible to determine quantitatively the amounts of various materials on a paper chromatogram. If the spots or bands are colored, the absorption of light by the materials will be related to their concentrations in the spot. If this method is not practical because of great differences in colors of materials or because they are uncolored, they may be stained with a suitable dye and then scanned with a modified colorimeter or spectrophotometer.

AMINO ACIDS Proteins are among the very most important biological compounds, and frequently the analysis of the amino acids composing a protein gives important information. The group of about twenty amino acids that result from the hydrolysis of a protein could be separated by any of the chromatographic methods.

Amino acids possess both acidic and basic groups in the molecules, and most have side chains that may be acidic, basic, or neutral. The individual molecules may have a net acidity or basicity and thus can be separated on ion exchange columns. Passage through a resin such as Dowex-50 might separate the acidic from the basic amino acids. Often a complete mixture of amino acids is passed through several columns in succession, each column containing a different resin. Careful control of the *pH* of the developing solvent permits the elution of virtually all the various amino acids as individual bands. Several models of instruments have been fabricated which will perform such a separation, determine the concentration of each constituent, and plot the results on a strip-chart recorder, all quite automatically!

For filter paper partition chromatography of amino acids, of all the different solvent mixtures that have been used, only a few have become popular: a mixture of phenol and water, a mixture of collidine, lutidine, and water, and a mixture of one of the butanols, with acetic or propionic acid, and water. Changes in operating conditions, modification of the proportions of the solvent materials, or use of two different mixtures makes two-dimensional chromatography possible. The resolved spots usually are detected and frequently are determined quantitatively by treating the developed chromatogram with ninhydrin (triketohydrindene hydrate). This treatment produces spots in various shades of pink or lavender.

CARBOHYDRATES   The sugars and sugar derivatives and the more elaborate polysaccharides are an extremely complex group. The biologist who must deal with these compounds is wise to enlist the help of an organic chemist. Carbohydrates often are modified chemically before chromatography and then can be separated by adsorption, ion exchange, or partition chromatography. Such a range of techniques is in use that it is difficult to say that one method is more commonly used than any other. Paper chromatography of simple sugars is performed on Whatman No. 1 or Schleicher and Schuell No. 589 White Ribbon paper, using the same solvents as for amino acids, or slight modifications thereof. The general reference books listed at the end of the chapter give fairly detailed descriptions of various methods and numerous citations of the original literature.

LIPIDS   The lipids include fats, waxes, and other materials that dissolve in any of a set of nonaqueous solvents. In the broad sense, the chloroplast pigments are lipids, and, in fact, the chromatographic methods used for lipids in general are only slight modifications of those used for the chloroplast pigments.

PROTEINS   The chromotographic separation of proteins, including the enzymes, has been difficult because the molecules are large and do not move freely, the molecular surfaces possess a wide range of chemical and electrical properties, and the proteins themselves are relatively unstable. Ion exchange methods seem to have produced the best results, but the behavior even on ion exchange columns probably involves some adsorption. Columns of tricalcium phosphate gel and of several Dowex and Amberlite resins are satisfactory for some proteins. A number of materials prepared by a chemical treatment of cellulose have been remarkably successful in the separation of a variety of proteins. These include carboxymethyl-cellulose and DEAE-cellulose

(diethylaminoethyl). ECTEOLA-cellulose (from epichloro-
hydrin and triethanolamine) is not as favorable for protein
but is a superior material for separating nucleic acids. Gel
filtration will probably become the most popular method for
chromatographic separation of proteins.

GAS CHROMATOGRAPHY A major development is gas chromatography, a separation
of materials in the vapor phase. A tube is packed with a solid
material which is then coated with a selected solvent. In
another system, the solvent forms a thin coating on the inner
walls of a long, very fine capillary tube. In either case, a
mixture of gases is allowed to pass through the tube. Those
components which are least attracted to the solvent appear
earliest at the far end of the tube, the other components
appearing later. As the individual gases emerge they are
detected by a device that measures thermal conductivity or
some similar physical property. The amount of each component is recorded electrically. So far, gas chromatography
has been more useful in chemistry and in industry than in
biology, mainly because most biological compounds cannot
be vaporized easily. Even so, the well equipped biology
laboratory has the requisite equipment. Probably it will be
possible to modify the method for the separation of a wider
range of biological compounds, as for example, by the
preparation of volatile derivatives of biological compounds
by various chemical procedures.

PROBLEM 1. Compare the effectiveness of paper, thin layer plates, and
columns in the separation of a mixture of biological materials.
The evaluation of the separations will be more convenient if
you select a mixture of colored materials.

SELECTED REFERENCES Several of the books listed in the bibliography contain sections
on chromatography.

Determann, H., *Gel Chromatography*. New York: Springer-
Verlag, 1968. Gel filtration, molecular sieves; a laboratory
handbook.

Heftmann, Erich, ed., *Chromatography,* 2nd ed. New York:
Reinhold Publishing Corporation, 1967. An encyclopedic

reference work containing articles on theoretical and practical chromatography written by experts in the field.

Stahl, Egon, ed., *Thin-Layer Chromatography*. New York: Springer-Verlag, 1965. The standard reference work in this field.

Journal of Chromatography, a periodical. Worth checking for specific methods.

Probably the single most valuable technique to become available to the biologist in recent years is the isotopic tracer method. Increased understandings in a number of important areas can be attributed directly to these materials, which came into general use shortly after the end of World War II.

The use of tracer isotopes is such a broad and highly technical subject that no attempt at complete coverage will be made here. Because the use of radioactive tracers, particularly, has potential dangers, I recommend that no one should amuse himself by simply playing with radioactive materials but instead should undertake tracer experiments only after competent instruction in the laboratory. Fortunately, although it is impossible to give more than a summary here, a number of excellent publications are available for the reader who is interested in pursuing the subject.

Several kinds of problems, otherwise insolvable, are experimentally easy if tracers are used. If a material moves from one place to another within an organism but several different pathways are possible, the tracer can identify the pathway taken. For example, mineral ions move from the roots where they are absorbed upward to the leaves of plants. They might move through either xylem or phloem; the proper application of tracer experimentation tells which tissue is the actual path. In an animal, a certain material might move from place to place through blood or lymph, and a tracer could be used here also.

Another major kind of problem is the chemical problem. We know that A is converted to Z, but paper chemistry tells us that any of several sequences of intermediates could be involved. A series of tracer experiments can delineate the chemical pathway. As a less extensive modification, a tracer can tell us whether a particular reaction occurs at all. If we provide cells with the suspected substrate of a reaction labeled with a tracer and then allow time for the reaction to occur, we should be able to recover the product of the reaction, now labeled with the tracer. More specific examples are given later.

Even whole cells, particularly microbial cells, or whole animals may be labeled with tracers.

THE TRACER EXPERIMENT The isotopic tracer experiment consists of substituting an unusual or uncommon isotope of an element for the more abundant forms. For example, radioactive carbon (^{14}C) can be substituted for ordinary carbon (^{12}C) and will go through the same chemical reactions. For the purposes of tracing, it is not at all necessary that every atom should be the unusual isotope. A small fraction of the tracer isotope serves to label the whole amount of a compound. The tracer experiment, then, requires an element or compound in which small fractions of the molecules contain the unusual isotope of an element. The other chief requirement in the experiment is some method for detecting the uncommon isotope before, during, and after the process being investigated. The fraction of the atoms which must be labeled depends upon the ability of the detecting method to distinguish the tracer atoms.

"Isotope" means literally "same place" and refers to atoms that occur at the same place in the periodic table; therefore, they contain the same number of protons and electrons, which determine the chemical activity, but, because they contain different numbers of neutrons, they differ in mass. The hydrogen series illustrates the point reasonably well. Ordinary hydrogen *is* ordinary hydrogen because it makes up about 99.98 per cent of the naturally occurring hydrogen. It is the simplest possible atom, consisting of one proton and one electron. About 0.02 per cent of the naturally occurring hydrogen atoms contain a neutron in addition to the proton and electron. This neutron doubles the mass of the atom without appreciably changing its chemical properties. A third type, present as a trace in nature but manufactured artificially in reactors, contains a second neutron and therefore has three times the mass of ordinary hydrogen. These three isotopes of hydrogen are designated ^{1}H, ^{2}H, ^{3}H. The superscript (1, 2, 3) indicates the mass. ^{2}H is frequently called deuterium and is sometimes given the symbol D, whereas ^{3}H is called tritium (T).

RADIOACTIVITY AND RADIATION The nuclear combination of one proton and one neutron is stable; that is, there is no tendency for this nucleus to decompose. The tritium nucleus, however, is unstable and undergoes spontaneous degradation to a more stable form. The excess energy of the unstable form is given off as a radioactive emission, in this case β^- particles or electrons, as one of the neutrons changes to a proton and an electron.

The atom becomes ^{3}He, and the electron or β^- particle is accelerated through space. Other radioactive materials disintegrate in this manner or in other patterns which yield α particles (2 protons + 2 neutrons), β^- (electrons) or β^+ (positrons), γ quanta (electromagnetic radiation), or some combination of these. Any isotope disintegrates at a characteristic rate in a characteristic way.

The α, β, and γ emanations possess great energy. As they move through air or other materials they produce pairs of ions. They share this quality with X rays, which are produced by electronic rather than nuclear events. Quanta of visible light and ultraviolet are basically similar to γ rays and X rays, but the quanta possess insufficient energy to produce sustained ionization. For this reason, only X rays and radioactive emanations are classified as ionizing radiation.

Amounts of radioactive materials are measured in curies, one curie being the amount of a radioactive material such that 3.7×10^{10} atoms disintegrate per second. A curie is a large amount of radioactive material, and biologists are more likely to work with millicuries (mc) or microcuries (μc). The basic unit of ionizing radiation is the roentgen (r). The roentgen is an amount of X or γ radiation sufficient to produce about 2×10^9 ion pairs in 1 cm^3 of dry air. Several other units of radiation have been developed for use in studies on the biological effects of radiation. These include rep (radiation equivalent, physical), used in measuring radiation absorbed by soft tissue; the rad (radiation, absorbed dose), the amount absorbed in any medium; and the rem (radiation equivalent, man) which applies specific corrections for man.

Ionizing radiation has profound effects upon all kinds of biological materials. The study of these phenomena is a gigantic field in itself, somewhat outside our purpose here. Thus, although the biological effects of radiation cannot be separated from the use of isotopic tracers, the reader must be referred to some of the works listed at the end of the chapter.

THE DEVELOPMENT OF TRACER EXPERIMENTATION

Tracer experiments were used for the first time by Hevesy in 1923. He used several naturally occurring radioactive isotopes of lead to trace the path by which materials moved from one place to another within plants. In 1934 the famous Curies demonstrated the possibility of producing artificial radioisotopes (a contraction of "radioactive isotopes"), and even before the beginning of World War II several isotopes had been produced artificially. Most of the potential tracers were not yet available in adequate quantities, however, so only a few isotopic tracer experiments were performed. After World War II great quantities of many different isotopes became quite readily available, and laboratories all over the world adopted this new tool. Some difficulties were encountered, and, in fact, some problems not particularly amenable to solution by the use of tracers were investigated. By now the "fad" has passed, and we have settled down to a judicious use of this most valuable technique. Tracer materials are now available in almost any conceivable form, and the instruments used for detection have reached a high state of development.

SELECTION OF TRACER ISOTOPES So many isotopes have been produced artificially and are potentially available that the investigator is faced with a choice of isotopes. For example, there are several isotopes of carbon, some stable, some radioactive. If a convenient radioactive isotope is available it usually is chosen as a tracer because the detection of radioactivity is easier than the detection of stable isotopes. Two principles are considered in selecting from among several possible radioisotopes: the rate of disintegration and the type of disintegration and radioactive emission.

Any radioactive isotope disintegrates at a characteristic rate. Any single atom has a certain probability of decomposing, regardless of how many similar atoms are in the vicinity. This probability means that in a unit of time a certain constant fraction of the total will disappear. Thus, $dN/dt = -KN$, where N is the number of atoms, t is time, and K is a constant representing the fraction of the total number of atoms disintegrating in a unit of time. From this equation it can be seen that half of the atoms will disintegrate in some certain time. The same length of time is required for half of the remaining atoms to disintegrate. This disintegration rate can be seen more easily in Fig. 12-1. The time required for half the atoms to disintegrate is called the half-life of the isotope, designated as $t_{1/2}$. Since the rate of disintegration of any isotope is a distinctive property of that isotope, each isotope has a characteristic half-life. Biologists have become so accustomed to the term that they now facetiously speak

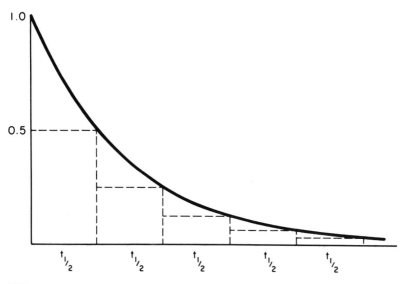

FIGURE 12-1   *Curve describing decay of a radioactive isotope.*

of the half-life of their paychecks and of other items to which the reasoning is not strictly applicable.

The half-lives of isotopes influence the selection of tracer isotopes. Obviously the half-life must be some convenient period of time. If half the isotope disintegrates in 0.01 sec, any experiment must be done in an impossibly short time. The only radioisotopes in common use as tracers have half-lives of at least several days; ^{14}C decays so slowly that half of it still remains after almost 6000 years. Thus in case of a choice among several radioisotopes of the same element, some always have more favorable half-lives than others.

A second major consideration in choosing radioisotopes is largely a matter of laboratory safety. Alpha-emitters are not commonly used as biological tracers. There is often a choice, however, between β- and γ-emitters, and in this case the β-emitter is likely to be chosen because the less-penetrating and less-energetic β particles are less dangerous to personnel.

As an example, let us examine the six isotopes of carbon: ^{10}C, ^{11}C, ^{12}C, ^{13}C, ^{14}C, and ^{15}C. ^{12}C is the abundant or normal isotope; ^{13}C is stable and occurs naturally (about 1 per cent); ^{14}C is radioactive, a β-emitter, and occurs naturally as traces. The radioisotopes ^{10}C, ^{11}C, ^{14}C, and ^{15}C have all

been produced artificially and have the characteristics listed
in Table 12-1. Carbon, of course, is the biological element,
if any element can so qualify. ^{14}C was the obvious choice
as a tracer because of its long half-life and because of its
relatively safe production of β^- particles. The first radio-
carbon tracer experiments were performed with ^{11}C because
it was produced earlier, but as soon as ^{14}C became abundant
it was immediately adopted.

TABLE 12-1 *radioactive isotopes of carbon*

isotope	half-life	type of emanation
^{10}C	19 *sec*	β^+, γ
^{11}C	20 *min*	β^+
^{14}C	5730 *years*	β^-
^{15}C	2.4 *sec*	β, γ

Other factors may influence the choice of isotope if more
than one element can be considered. In a given experiment
it might make little difference whether the substrate molecules
were labeled in the carbon atoms or in the hydrogen atoms.
There might be very good reasons for choosing tritium rather
than ^{14}C as the tracer. Tritium has a very high specific
activity, i.e. a large amount of radiated energy per unit of
atoms. This means that the tracer can be diluted considerably
in the course of the experiment and still be detected. As one
example, thymidine, one of the building blocks of DNA, can
be obtained labeled with ^3H. If tritiated thymidine is provided
to a multiplying virus, each new virus particle will be
detectably tagged, even though each may contain very little
tritium.

**COMMONLY
USED TRACERS**   Biologists usually deal with relatively few of the elements,
chiefly those in the lower one-third to one-half of the periodic
table. Most of these elements have at least one convenient
radioisotope. The important exceptions are nitrogen and
oxygen, both of which are extremely important in biology.
Table 12-2 lists the isotopes most commonly used as tracers
in biology.

In addition to these commonly used tracers, several other radioisotopes are important in biology. ^{60}Co is used frequently as a γ source for experiments on the effects of ionizing radiation. Masses of ^{60}Co up to several hundred curies can be used in properly shielded pieces of apparatus into which biological or chemical materials can be introduced for varying lengths of time. A cesium isotope (^{137}Cs) is used for similar purposes, offering easier handling but somewhat less energetic radiation. Radioactive rubidium (^{87}Rb) is sometimes used as a tracer, not for ^{85}Rb but for potassium, which it resembles

TABLE 12-2   *some commonly used tracer isotopes* *

"normal" isotope	*common tracer*	*half-life*	*emanation*
^{1}H	^{2}H		*stable*
	^{3}H	12.3 *yr*	β
^{12}C	^{13}C		*stable*
	^{14}C	5730 *yr*	β
^{14}N	^{15}N		*stable*
^{16}O	^{18}O		*stable*
^{23}Na	^{22}Na	2.6 *yr*	β, γ
^{31}P	^{32}P	14 *days*	β
^{32}S	^{35}S	87 *days*	β
$^{35, 37}$Cl	^{36}Cl	3×10^5 *yr*	β
^{39}K	^{40}K	1.2×10^9 *yr*	β, γ
^{40}Ca	^{45}Ca	165 *days*	β
^{56}Fe	^{59}Fe	45 *days*	β, γ
^{59}Co	^{60}Co	5.2 *yr*	β, γ
^{127}I	^{131}I	8 *days*	β, γ

* Other isotopes of sodium, iron, cobalt, and iodine are available and are sometimes used as tracers.

chemically. Some cells accumulate Rb almost as readily as K, so the Rb can be valuable in studies on cell membrane permeability. ^{90}Sr is also important biologically because of its chemical similarity to another element, calcium. Several biological mechanisms tend to concentrate the ^{90}Sr which might occur in fall-out from nuclear weapons. Iodine collects in the thyroid; therefore ^{131}I has been used, not only as a tracer, but for radiation therapy in disorders of the thyroid.

AVAILABLE
FORMS OF
TRACER MATERIAL

Radioisotopes formerly were available chiefly as some salt containing the tracer element. ^{14}C, for example, was delivered as $Ba^{14}CO_3$. The carbonate could be converted to $^{14}CO_2$ by treatment with acid and ultimately converted into any of several chemical compounds. Now producers of chemical and biochemical compounds offer long lists of organic chemicals labeled with any of several tracer isotopes. It is even possible to specify which atom (or atoms) will be labeled. Acetic acid, for example, can be purchased as $^{14}CH_3{}^{12}COOH$, $^{12}CH_3{}^{14}COOH$, or $^{14}CH_3{}^{14}COOH$. Most of the sugars, most of the amino acids, and a variety of other compounds are available, labeled with ^2H, ^3H, ^{13}C, ^{14}C, ^{15}N, ^{18}O, or assorted other tracers. The availability of these compounds considerably simplifies the execution of tracer experiments.

DETECTION
METHODS

Tracers, of course, are useless unless they can be detected after the experiment is complete. Since very small amounts of tracer materials are used, special methods are necessary to recover the tracer. A number of methods have been used to prepare materials for examination. If a physical movement from place to place is being investigated, nothing more than a dissection of the animal or plant may be required. Chemical experiments offer more difficulty. The suspected product of a reaction must be separated from other compounds in the cells or reaction mixtures; nowadays chromatography is used for this separation.

AUTO-
RADIOGRAPHY

One simple method of detecting radioactive materials is to place the plant or animal parts—or the chromatogram— against a sheet of photographic film. The radioisotopes produce an "autoradiograph" because the irradiated areas of the film show up on development. Such a technique would almost always be used to find *where* the tracer is located rather than how much is present. Rough quantitative estimates can be obtained by measuring the relative darkening or density of the film. More precise determinations require the very tedious counting of silver grains under a microscope.

Ordinary photographic film on a cellulose ester base is used for chromatograms and other relatively large objects. For example, a leaf suspected of containing radioactive materials is wrapped in thin plastic film (Saran, HandiWrap) to protect the emulsion from leaf juices. The wrapped leaf is then clamped against a sheet of photographic film such as would be used in diagnostic X-ray machines.

This whole operation goes on in the dark, of course, or else the photographic film must be kept wrapped in black paper.

On a microscopic scale it is more difficult to use regular commercial film. Photographic emulsions on some novel bases are now made especially for autoradiography of small specimens. A section of plant or animal tissue on a microscope slide is covered by a thin protective coating of inert material. The photographic emulsion, without the usual backing, is placed in intimate contact with the biological material. Stripping film, a preparation in which the protective and supporting backing peels off from the emulsion, is most commonly used. The flexible emulsion layer can conform to the contours of the material on the slide. Liquid emulsions, which dry rapidly to form a coat of sensitive material, are also used. After a suitable period for exposure (in the dark, of course), the microscope slide, with the emulsion adhering to it, is developed in the usual manner. Such "micro" auto-radiograms can show the accumulation of tritium in the individual chromosomes of cells.

THE G-M TUBE   A number of electrical instruments are employed for quantitative determination of radioactivity. The Geiger-Müller (G-M) tube is probably still most commonly used. It consists of a hollow metal tube, filled with a gas mixture, with a wire extending along its center for most of its length. The circuitry is diagrammed in Fig. 12-2. The behavior of the tube depends on the applied voltage. Over the commonly used range of voltages, the so-called G-M region, the tube responds to incoming radiation in a characteristic way. (At lower voltages, different but frequently useful behaviors are observed.) Beta particles enter through the window, gamma quanta through the window or walls, and produce ions in the gas in the tube. At this voltage, negatively charged ions migrate toward the center wire, producing other ions as they travel. Eventually a "cloud" of ions strikes the center wire as a pulse of charged particles, setting up a momentary electric current in the wire. Positively charged ions move in the opposite direction, producing the same effect. Electric neutrality is re-established in the gas, and the tube is ready

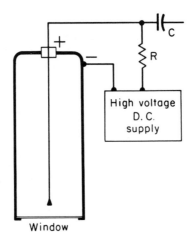

FIGURE 12-2   *A Geiger-Müller tube. The window is frequently of mica.*

to accept another particle or quantum. The pulse of current crosses the capacitor to be registered on a meter or on a scaling or counting circuit. The whole operation is completed in about a microsecond or so. If a radioactive sample is placed under the tube, some constant fraction of the radioactivity will enter the tube to be counted, the exact fraction depending on the geometry of the system. Amounts of radioactivity measured this way are usually expressed as counts per minute (c/m).

THE SCINTILLATION COUNTER

This instrument is more easily adapted to special measurements and has become increasingly popular in recent years. The scintillator is a material which responds to radioactivity by producing flashes of light. These individual flashes are counted by a multiplier phototube (Chapter 13); the electrical output of the phototube is counted by a scaling circuit. The scintillating phosphor may be dissolved in a liquid, which means that the counting chamber can take almost any shape. Whole-body counters have been constructed as hollow cylinders in which dogs or other animals are placed. The scintillation counter surrounds the body and detects any radioactivity emitted from the animal's body.

In the more common laboratory usage, chemical compounds suspected of containing radioisotopes are placed in small glass or plastic vials and carefully dried. Scintillation fluid, a solution of an organic phosphor in toluene, is added, the vial is capped, and is then placed in the counter.

Different isotopes produce emanations of differing energies. The scintillator-phototube system detects these differences in energy. Therefore, with suitable electronic circuitry in the scaling system, it becomes possible to distinguish among various isotopes. For instance, ^{14}C and ^{3}H can be counted in the same sample.

SCALING CIRCUITS   The output from a G-M tube or a scintillation counter is a series of electric pulses. The rate at which these pulses are produced can be measured with an ammeter calibrated in amperes or in c/m. Alternatively, each pulse can be counted singly with a digital computing circuit known as a scaler. The electronic scaler can be arranged in a variety of ways to indicate a total number of counts in a certain length of time or the length of time required to reach some predetermined number of counts.

Counts of radioactive materials always are accompanied by determinations of "background," that is, pulses produced from uncontrollable sources such as cosmic rays or γ-emitters in the general vicinity of the counting tube. The background count, assumed to be constant during the measurement, is subtracted from the total number of counts. Since background varies from day to day, each measurement must be corrected for background. Background can be reduced by shielding the counting tube with lead or by rather intricate electronic correction, but it can never be eliminated.

ACCESSORY   Samples of materials to be counted are placed in small metal
EQUIPMENT   dishes and placed beneath the window of the G-M tube. The samples are counted for a period of time adequate to obtain a good measurement of average rate of radioactive disintegration, and then another sample can be counted. Several automatic sample changers are available. These devices place one sample under the tube, count to a predetermined number of counts, print the time required, and then place another sample under the G-M tube. Thirty-six or so samples can be counted in sequence over a period of several hours without any attention from the operator.

Radioactive materials on paper-strip chromatograms are counted by strip scanners which automatically feed the strip of paper under the counting tube and then print or record the radioactivity as a function of distance along the paper strip.

Laboratory scintillation counters incorporate sample changers also. The small vials are placed in a conveyor chain and, one after another, are lowered by an elevator into the counting chamber in close proximity with the phototubes. Counting starts automatically and continues for a preset time or to a predetermined number of counts, at which time the next vial is moved into place. The results are printed on a paper tape, or in the more expensive models, by an electric typewriter. The sample changers hold as many as 200 small vials. The instrument is frequently allowed to run without attention overnight.

DETECTION OF
STABLE ISOTOPES

As stable isotopes differ from each other only in mass, any measurement of these isotopes must depend on this property. A mass spectrometer is an instrument which separates molecules on the basis of mass and determines the amounts of each kind of molecule. The mass spectrometer consists of an evacuated tube across which a high voltage is applied. The gas to be analyzed is admitted at one end of the tube, ionized by a stream of electrons, and accelerated toward the opposite end of the tube. Figure 12-3 shows one such instrument, and Fig. 12-4 shows another which separates materials on a different principle.

In the "time-of-flight" mass spectrometer (Fig. 12-3), ions traverse the length of the tube, but the greater the mass, the longer this passage will take. As the ions arrive at the negative electrode they set up a momentary current whose magnitude is dependent upon the number of such ions. If the tube is long enough, ions of different masses arrive at distinctly different times. Once all the desired masses have been measured, a new quantity of gas can be admitted and the operation repeated. The repetitions can be extremely rapid (up to several thousand cycles per second), so that the "mass spectrum" can be displayed on a cathode-ray oscilloscope.

The other basic type of mass spectrometer (Fig. 12-4) separates ions of different mass in a magnetic field. The tube has a 60°, 90°, or 180° bend surrounded by the poles of a large magnet. As the ions traverse the tube they are deflected by the magnetic field through an angle dependent on their

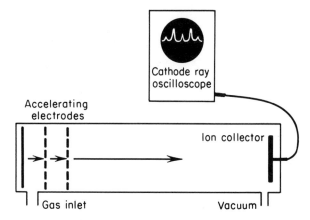

FIGURE 12-3   *A time-of-flight mass spectrometer. Gas is admitted, ionized by a stream of electrons, and accelerated toward the ion collector by the accelerating electrodes. Current on the ion collector is registered on the oscilloscope.*

momentum. The accelerating voltage and the magnetic field can be adjusted to focus ions of a selected mass on the target electrode. By varying the accelerating voltage or the magnetic field or both, the whole spectrum of ions can be swept across the target. Current in this electrode will vary according to the number of ions, just as in the time-of-flight instrument. If the "mass spectrum" is scanned automatically, the instrument will record the relative amount of each component in the mixture.

The actual operations required in a tracer experiment using stable isotopes become somewhat complex. We might perform a tracer experiment in which ^{15}N is used to follow the production of ammonia (NH_3). $^{14}NH_3$ has a mass of 17, but $^{15}NH_3$ has a mass of 18, and the two could be separated on the mass spectrometer. Ordinary water also has a mass of 18, however, and H_2O would obscure the $^{15}NH_3$. In this case it would be preferable to convert the nitrogen to some other form with unambiguous mass numbers. If we measure $^{12}CO_2$ (Mass 44) and $^{13}CO_2$ (Mass 45), it becomes necessary to correct for ^{17}O ($^{12}C^{16}O^{17}O$: Mass 45). Except for these technical details which must be considered, the mass spectrometer permits tracer experiments with stable isotopes, experiments which can be just as effective as radioisotope experiments.

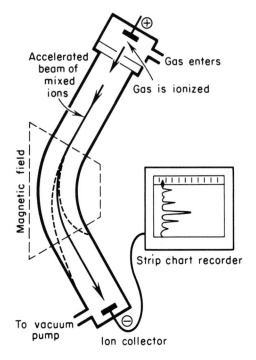

Accelerated beam of mixed ions

Gas enters

Gas is ionized

Magnetic field

Strip chart recorder

To vacuum pump

Ion collector

FIGURE 12-4   *A magnetic field mass spectrometer. Ionized gas is resolved by the magnetic field into a spectrum of particles of different sizes.*

LABORATORY SAFETY

Most radioactive materials used as tracers are by-products of reactor operations controlled by the United States Atomic Energy Commission. The AEC specifies the conditions under which radioactive materials may be handled. Very small quantities (μc or fractions of μc) of a large number of isotopes can be purchased by any citizen under a "General License." Usually these amounts are so small that serious radiation danger is unlikely even with somewhat careless handling. The amounts are adequate for certain kinds of experiments, however.

Larger amounts of isotopes, as would be used in a typical research laboratory, must be procured under a special license granted to the laboratory by the AEC. The license specifies the kinds and amounts of isotopes that can be held in the laboratory at any one time. One person is usually designated as Radiation Safety Officer, and he has the responsibility for maintaining careful records, supervising handling and disposal of radioactive materials, and protecting laboratory personnel against harmful exposure to radiation. No person or laboratory is granted such a special license unless the AEC is given the assurance that only persons trained by course work or "on-the-job" experience will be responsible for the handling of radioactive materials.

SELECTED
REFERENCES

Chase, Grafton D. and Joseph L. Rabinowitz, *Principles of Radioisotope Methodology,* 3rd ed. Minneapolis: Burgess Publishing Company, 1967. One of the standard reference works in the field; includes many experiments.

Gude, William D., *Autoradiographic Techniques.* Englewood Cliffs, N. J.: Prentice-Hall, Inc., 1968.

Kamen, Martin D., *A Tracer Experiment: Tracing Bio- chemical Reactions with Radioisotopes.* New York: Holt, Rinehart and Winston, Inc., 1964. A single experiment; theory, methods, interpretation.

Wang, C. H., and David L. Willis, *Radiotracer Methodology.* Englewood Cliffs, N. J.: Prentice-Hall, Inc., 1965. Another of the standard works.

Many biologists prefer to make their measurements electrically whenever it is possible. Much of the information obtained in biological experiments is indirect information anyway. If the possibility exists of converting this indirect information into an electrical signal of some sort, several advantages accrue. The measurement can be made more-or-less automatically, which often reduces the chances of human error. The electrical devices can produce permanent records, as on strips of chart paper, and these records can be examined and re-examined as needed. The electrical instruments usually respond very rapidly, so some responses too fast for the human senses can be detected easily. Electrical quantities are rather easily converted from one form to another.

Certainly there are disadvantages, too. Electrical and electronic equipment can fail, and sometimes does at the most awkward moments. The operator must be able to recognize faulty performance and, ideally, should know what to do about it. The electrical instrument measures only indirectly, necessitating the assumption that the electrical signal is some unvarying mathematical function of the biological response. Electrical instruments usually are expensive and require maintenance. In many kinds of experiments, however, the advantages outweigh the disadvantages.

**ELECTRICAL
THEORY**

Current electricity, as opposed to static electricity, is used almost exclusively in electrical instrumentation. This fact somewhat simplifies the explanations. A current can be thought of as a stream of electrons moving through a conductor, even if it is unlikely that any one electron travels very far. Material such as metals, in which certain electrons are rather loosely bound to the atom, make good conductors because the electrons can move rather easily from one atom to another. A variety of other materials in which the electrons have almost no freedom to move conduct very poorly. Intermediate between these two extremes are a group of "semiconducting" materials.

If electrons are to move primarily in one direction, and not at random, some force must be applied. This electromotive force is a difference in electrical potential, measured in volts. The ability of a material to conduct electrons is usually expressed by the inverse property, or resistance. The resistance of a wire depends upon its cross-sectional area, its length, and the metal of which it is made, and can be varied by changes in temperature or mechanical stress.

Voltage (E), current (I) and resistance (R) are related through Ohm's law, $I = E/R$. If we know any two of these quantities, we can calculate the third. Another quantity, electrical power, is the product of voltage and current $(W = IE)$ and is measured in watts. A null-type measurement in common use for electrical quantities employs a Wheatstone bridge (Figure 13-1). A pair of matched resistors is connected, as shown, with another pair of resistors, and a voltage is applied from the battery. If the resistance of R_3 is greater than R_4, a current in one direction is indicated by the meter; or if R_4 is greater than R_3, current will flow in the opposite direction. The resistor R_3 can be adjusted until there is no current so that we can measure R_4 in terms of R_3. Because voltage, current, and resistance are precisely interrelated by Ohm's law, the basic bridge circuit can be adapted to a variety of measurements where a third "unknown" value is calculated from the other two.

A current of electrons sometimes flows through a conductor in a single direction. In these direct current (dc) situations, Ohm's law applies without modification. Alternating currents, to the contrary, change direction with regularity and periodicity. The sine curve is derived from the path traced by a point on the circumference of a circle. At any one instant the voltage is positive, negative, or zero; current will flow in a circuit in a forward direction, in the backward direction, or not at all. Over one cycle, the voltage rises from zero to a maximal positive value, falls to zero, then rises to a maximal negative value, and again falls to zero (see Fig. 13-2). Alternating currents can perform work, just as direct currents can, and are characterized by a frequency.

The alternation of the current introduces some interesting and valuable complexities into the consideration of electricity. Any conductor carrying a current is surrounded by a magnetic field. When the circuit is broken this field collapses, only to expand again when the current starts flowing. Any dc conductor, then, is surrounded by a steady magnetic field which goes through transitory changes only when a switch is opened or closed. The ac conductor, however, is surrounded by a magnetic field which is continually expanding or collapsing. If this expanding and contracting magnetic field moves through a second wire, a

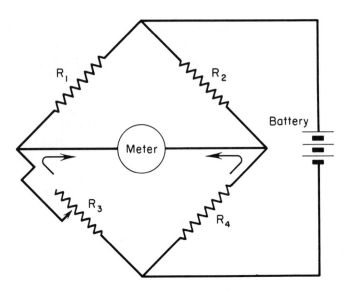

FIGURE 13-1   *"Null measurement" of resistance is performed with this Wheat-stone bridge. $R_1 = R_2$, R_3 is adjustable and of known resistance, R_4 is unknown. When $R_3 = R_4$, zero current is shown on the meter.*

current will be *induced* in this second wire. Now imagine a coil of wire conducting an alternating current. The expanding magnetic field around one turn of the coil cuts across the next turn in the coil, inducing a current in this second turn of the coil *in the opposite direction*. The net effect throughout the coil is an induced current of opposite sign, which tends to impede the original current. Impedance from this source occurs in addition to the regular resistance of the wire in the coil. Thus in any treatment of an ac circuit which contains such coils we must consider the inductance as well as the resistance.

Another device produces a more-or-less opposite effect. A capacitor (condenser) consists of a pair of plates of conducting material separated from each other by a dielectric (insulating) material. If the two plates are connected by wires to a dc circuit, no current will flow through the dielectric material, but an excess of electrons accumulates in one of the plates and a deficiency in the other. The capacitor becomes charged, one plate positively, the other negatively. In an alternating circuit, electrons pour freely into one plate,

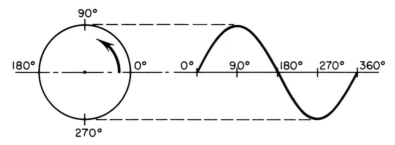

FIGURE 13-2   *Relationship of the sine curve to the circle.*

forcing electrons out of the other plate. When the sign of the current is reversed, the events occur in the opposite direction. The capacitor seems to conduct ac. The current flows so freely into the large plates of the capacitor, however, that the resistance seems smaller than if the capacitor were not present.

The inductive effect of a coil and the capacitive effect of the condenser become very important in any ac circuit in which they occur. Instead of ordinary resistance, we speak of impedance (Z), and $I = E/Z$. The inductance (L) of a coil (in henries) affects the current in a manner which is dependent on the alternating frequency (f). The opposition to the current is called inductive reactance, $X_L = 2\pi fL$. The condenser of capacitance C (in farads) produces an opposite effect, and capacitive reactance $X_C = 1/(2\pi fC)$. The impedance Z is the vectorial sum of resistance R and the net reactance.

$$Z = \sqrt{R^2 + (2\pi fL - \frac{1}{2\pi fC})^2}$$

In any ac circuit containing both inductance and capacitance the actual current flowing under a given voltage could be calculated from this equation. Impedance defined in this manner becomes particularly important in communications where frequencies are high. The ideas are used in biological instrumentation, of course, and should not be neglected even in measuring electrical behavior of tissues. Living materials possess electrical resistance, but there is also likely to be a measurable capacitance within the living material, and another capacitance often exists in the connection between the electronic instrument and the living material.

VACUUM TUBES   Vacuum tubes are still used in a majority of instruments for a variety of purposes. A very large number of different tubes is available commercially, but in general they all depend on the same principles. If a conductor with its loosely bound electrons is heated, some of these electrons will escape from the surface. In a sense, they are "boiled off." In an ordinary conductor, however, the electrons will be immediately recaptured by the positive charges left on the surface of the metal. At any instant, the hot surface will be surrounded by a "space charge," a cloud of free electrons. If such a hot electrode is placed in a vacuum, together with another "cold" positively charged electrode, electrons leave the hot surface, travel across the intervening space, and are captured by the positively charged electrode (called the *plate* or *anode*). A "diode" operates in this way. (Fig. 13-3a shows such a tube.) Obviously the tube must be connected into a proper external circuit so that the hot *cathode* does not become positively charged by the loss of electrons. Also obviously, the stream of electrons can move in only one direction. Such a device could be used to rectify, that is, to change alternating current into direct current.

If a grid or screen of fine wires is placed between the cathode and the plate (as in Fig. 13-3b) and this grid is made slightly positive or negative, the small charge will interfere with the free movement of electrons across the space. Variations in the grid voltage are reflected in the current passing through the vacuum tube to the plate. The cathode-to-plate voltage is much higher than the grid voltage, so any "signal" or variation in grid voltage is thus amplified. The "triode" amplifier tube, with all the variations and improvements which have been made, is responsible almost by itself for our whole electronic world. Sometimes other electrodes are introduced, as in tetrodes and pentodes, but usually these are included to improve the performance of the basic triode.

Semiconducting devices, or transistors, can perform some operations similar to those performed by vacuum tubes. These solid state materials differ only in degree from conducting materials (metals) or from insulating materials. Crystals of germanium or silicon possess a definitely ordered structure in which pairs of electrons are shared by adjacent atoms. Traces of impurities may fit into the crystal structure but introduce extra electrons which are not needed in the

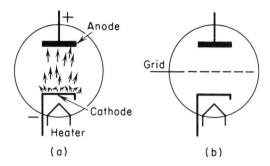

FIGURE 13-3   *Diode (a) and triode (b) vacuum tubes.*

crystal bonds and thus are free to migrate. Other impurities
may fit the crystal pattern, but with deficiencies of electrons
or "holes." Such "slightly impure" crystals become the basis
of transistors. A crystal of germanium (4 valence electrons)
with a trace of arsenic or phosphorus (5 valence electrons)
possesses an excess of electrons and is called N-type (for
negative) material. A similar crystal containing traces of
gallium or indium (3 valence electrons) would be called
P-type from its excess of "holes" or positive charges. If a
piece of N-type and a piece of P-type crystal are joined and
then a voltage is applied, electrons flow in one direction more
easily than in the other. "Holes" flow more easily in the
opposite direction. This N-P junction becomes a rectifier
(see Fig. 13-4a).

A combination of N-P-N layers is a junction transistor,
which can best be explained with the help of Fig. 13-4b. The
size of the current flowing across the N-P and P-N junctions
depends upon a number of factors. Electrons cross easily
from the negatively-charged emitter to the P-type base, filling
some of the holes in the base. The positively-charged collector
withdraws electrons from the base, across the P-N junction,
creating new holes. These new holes migrate across the base,
eventually being filled with electrons from the emitter. The
number of new holes formed—and therefore the current
flowing in the base-collector half of the transistor—depends
upon the number of electrons injected into the base from
the emitter. The transistor can thus accept a signal and
control the current in a second circuit. If the voltage in the
base-collector circuit is larger than that in the emitter-base
circuit, the control amounts to an amplification.

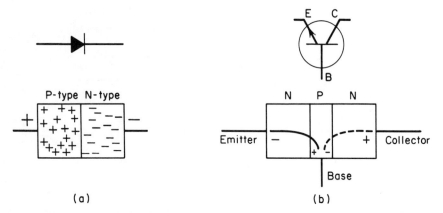

FIGURE 13-4   *Solid state (semiconducting) devices: (a), a junction diode; (b), N-P-N transistor. For explanation, see text.*

In practice, the N-P-N transistor can be used in a circuit in any of several ways. The P-N-P transistor is the same kind of device except that the polarity is reversed—positive poles becoming negative and vice versa—so that the current carriers through the base are electrons rather than holes.

Transistors have several decided advantages over vacuum tubes. Since they do not depend upon heaters, they can be operated with very little applied power. They also can be made very tiny, which helps to reduce the size and weight of instruments. Generally, they have a much longer life than vacuum tubes. Other advantages, and some disadvantages, will be discussed later under Amplifiers.

ELECTRONIC SYSTEMS   An electronic system, as the term is used here, is a device or combination of devices that responds to some change in the environment in a characteristic way to produce an electrical change in a measuring instrument or to bring about some control over the environmental change. Systems of this sort are adaptable to measurement of a variety of biological phenomena. Some biological reactions produce voltages directly, and these are relatively easy to measure. In other cases, the biological process can lead to a change in resistance or capacitance or some other property of a circuit. Movement can be detected by a mechanical coupling to an instrument which produces an electrical "signal." The modifications, interconversions, and variations are almost unlimited.

The biological response, or the response of a physical instrument used in a biological experiment, must be converted into some kind of electrical signal. Any device which does this is called an input transducer. It accepts information and converts it into some usable electrical form. In the simplest system the input transducer feeds directly into an output transducer, such as a meter. More commonly, the electrical signal is amplified. The system then becomes more complex because the amplifier, and perhaps the other components, require a power supply to furnish a number of dc and ac voltages. The output of the system may be recorded, or it may be used to control, or both. A complete system is diagrammed in Fig. 13-5. Each major component is considered in more detail in a succeeding section.

INPUT TRANSDUCERS

Input transducers which respond to a variety of signals are available. These are admirably covered in the book by Lion.* It is a shame that the other components of systems are not yet treated so well.

MECHANICAL TRANSDUCERS

Input transducers in this class respond to movements, changes in pressure, or other mechanical changes. A simple kind of motion detector is a variable resistor. Any movement of the movable contact along the "slidewire" changes the resistance in the circuit, and this change becomes the electrical signal. Figure 13-6 shows several such devices which respond to linear motion, rotary motion, and pressure.

A mechanical transducer could also produce changes in inductance or in capacitance. For example, the capacitance of a condenser depends, among other things, on the distance between the two plates. The Beckman InfraRed Analyzer (Chapter 10) employs a variable capacitance detector. Some record-player pickup arms use variable inductances. In both inductance and capacitance transducers, a moderately high-frequency "carrier" current is varied by the signal because both inductances and capacitances have their greatest effect

* Kurt S. Lion, *Instrumentation in Scientific Research: Electrical Input Transducers* (New York: McGraw-Hill Book Company, Inc., 1959).

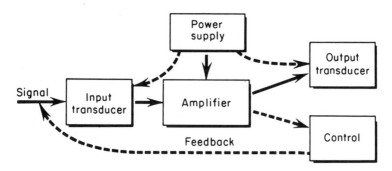

FIGURE 13-5   *A typical electronic system.*

at higher frequencies. The variations in "carrier" current then can be amplified.

Certain crystalline materials, quartz for example, exhibit the "piezo-electric" effect. If a voltage is impressed across a wafer of the crystal, the crystal changes its shape. If an alternating current is used, the crystal vibrates, vibrating most strongly when the frequency corresponds to the natural period of the crystal. This effect is used most commonly in regulating the frequencies in communications circuits. The reaction is reversible, however, making it possible to use such crystals as transducers. If the crystal is compressed or vibrated, it will generate an alternating voltage. This signal alternates with a frequency corresponding to the vibration of the crystal, so the device can be used to detect small motions.

Probably the most spectacular of the mechanical trans-ducers are the "strain gages." These devices depend upon the fact that the resistance of a wire changes a tiny amount when the wire is stretched. The wire (or a flat strip of a conducting metal) is arranged so that the pulling force is exerted on it, and at the same time, the wire is part of a Wheatstone bridge. Changes in resistance can be amplified and used to drive a recorder.

TEMPERATURE TRANSDUCERS   The thermocouple, the resistance thermometer, and the thermistor were discussed in Chapter 4. Each of these is properly called an "input transducer." The thermocouple produces a dc voltage directly, while the other two instru-ments give variations in resistance. In either case, it is not difficult to fit the transducer into an electrical system.

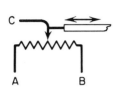 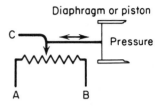

FIGURE 13-6   *Mechanical transducers in which resistance varies in response to linear motion (left), rotary motion of shaft (center), or pressure (right).*

RADIATION
TRANSDUCERS

Probably transducers which respond to radiant energy occur in the greatest variety. This is true partly because light is energy, and within the visible and ultraviolet range the quanta are large enough to cause measurable electrical or chemical effects. The photovoltaic cell, or "barrier-layer cell," is a semiconducting device. Quanta of light displace electrons within the crystal. With a proper arrangement of semiconducting and conducting materials, the continued displacement of electrons becomes a small current which can be measured. Voltages always are low, but the currents are large enough to be measured on sensitive meters. Many of the photoelectric exposure meters used in photography are barrier-layer cells.

A phototube is a vacuum tube in which the cathode is sensitive to light. Quanta of radiant energy impinging on the cathode release electrons which are drawn to the plate or anode. Since the current across the tube is dependent on the number of quanta striking the photoemissive surface in a unit time, the output of the tube is proportional to the light intensity.

A multiplier phototube is a phototube with a built-in amplifier called an electron multiplier. A series of dynodes (electrodes with increasing positive charge) is arranged so that electrons liberated from the photoemissive cathode are drawn to the first dynode. Each electron causes the emission of additional electrons from the dynode surface. As shown in Fig. 13-7, a combination of eight or nine dynodes can result in a considerable multiplication of the current. Multiplier phototubes respond rapidly to very dim light or to quite small changes in light intensity. These tubes have been incor-

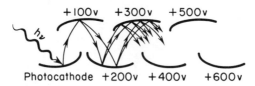

+100v   +300v   +500v

Photocathode +200v +400v   +600v

FIGURE 13-7   *Multiplier Phototube. A quantum of radiant energy causes the emission of an electron from the photocathode. At the first dynode (+100v) and at succeeding dynodes, additional electrons are liberated by secondary emission.*

porated in a number of standard laboratory instruments such as spectrophotometers.

Several semiconducting materials can be used effectively as light detectors because of a large decrease in resistance upon illumination. A dozen or more materials are available. Lead sulfide is used to measure infrared radiation in several of the commercial spectrophotometers. A typical cadmium selenide photoconductive cell is about the size of a small pea. These transducers have some disadvantages, but for certain applications in experimental research they can be very useful. Many semiconducting devices are now used in photographic light meters or inside the camera itself.

COMBINATIONS   In many instances, some combination of transducers can be used. Although a very small movement might be detected directly with a strain gage, it might sometimes be more convenient to measure such a movement by observing its effect on a beam of light. A sensitive galvanometer responds to a small current by moving a suspended coil of wire within a magnetic field. A small mirror, rotated along with the coil, reflects a beam of light which moves across the scale as the coil turns. The small mirror has less inertia than the needle in an ordinary meter, so the galvanometer responds to smaller currents. Thermocouples respond to temperature differences, but if one of the junctions is blackened it will absorb light and become warmer and thus will measure radiant energy. A scintillation counter is a derived instrument because ionizing radiation causes flashes of light in the phosphor, and these flashes are detected by the multiplier phototube.

OUTPUT TRANSDUCERS   Of the many types of output transducers, the galvanometer was the earliest used. If a coil of wire is suspended between the poles of a magnet, any current in the wire tends to deflect the coil transversely across the magnetic field. A typical measuring meter consists of a coil mounted on jeweled pivots surrounded by the magnet. The torque or rotating force in any given meter is proportional to the current through the coil. An attached pointer moves across a scale, and current can be read directly in amperes, milliamperes (mA), or microamperes (μA). The same meter can be used to measure voltage if an unvarying resistor is connected in series with the coil. Current through the resistor (and the coil) depends directly on the voltage. If the resistor is built into the instrument, the meter scale can be graduated in volts. A dry-cell (or other constant voltage source) and a meter can be used to measure resistance, once against because $I = E/R$.

SPEAKERS   The use of speakers as output transducers is sometimes advantageous. In a complex instrumental arrangement, for example, the operator's eyes might be too busy to watch a meter. His ears, then, could detect changes in volume or pitch from the speaker. In some instruments speakers are used chiefly for demonstration purposes, while in others the speaker gives a warning when some misuse of the instrument or some other disaster is imminent. Speakers can also be used in conjunction with other output transducers.

RECORDING INSTRUMENTS   The ultimate in convenience comes with the use of a recording instrument as the output transducer. A pen scribes a permanent record on a moving paper. Most frequently used are the various strip-chart recorders, in which paper from a roll is fed under the pen. The pen moves across the paper by an amount proportional to the strength of the electrical signal.

The recorder often contains a galvanometer and thus will measure current or, with an appropriate resistor, voltage. A pen is attached to the moving coil of the meter. Variations in the electrical signal are recorded as a curve, varying distances from some zero line. Since the paper moves with constant speed, the curve is plotted as a function of time.

The potentiometer recorder is more complicated but can respond more reliably to smaller fluctuations. The instrument is a null measuring device because the signal creates an

imbalance in the instrument. The instrument responds to
this imbalance by actuating a motor which increases or
decreases the resistance in the circuit, thus restoring the
balance. The amount of motor movement required to return
the instrument to the balanced condition is recorded on the
paper. In several of the instruments the pen is attached to
the same motor by a gear or pulley arrangement. The Brown
Electronik Potentiometer Recorder is one of the favored
instruments in laboratories everywhere. Several others may
be as good or even better for some purposes.

OSCILLOSCOPES   For relatively fast, high frequency responses, the cathode
ray oscilloscope is used. This instrument is a vacuum tube
in which one end is coated with a phosphor that glows
under the impact of a beam of electrons. At the other end
of the tube an electron "gun" beams electrons toward the
face of the tube. Along the electron path two pairs of charged
plates deflect the electron beam in the vertical or horizontal
direction. A potential difference between the vertical control
electrodes moves the electron beam upward or downward.
A similar pair of electrodes controls movement in the horizontal direction. In the laboratory oscilloscope the horizontal
control electrodes are set to make the electron beam sweep
across the face of the tube, return, and sweep again. The
sweep frequency can be a few hundred to many thousands
of cycles per second. The signal to be observed controls the
vertical movement of the electron beam. The phosphor on
the tube face glows for a short time after the passage of the
electron beam, so that in effect the face of the tube shows a
continuous curve, horizontally across the face, varying in
height as the signal varies.

The cathode ray oscilloscope is the standard test instrument
in the electronics laboratory and can be used for a variety
of other purposes.

COMPUTERS   As more and more laboratories have gained access to large
computers it has become possible to feed the output of one
or more instruments directly to the computer. Any calculations, including systematic corrections and statistical
computations, can thus be completed almost instantaneously.
The results might be printed out, or used to control some
phenomenon, or both.

POWER SUPPLIES   The electricity delivered to a building is almost always alternating current. In the United States it is 60-cycle alternating current, usually 110 to 130 volts or some multiple thereof. Except in special circumstances, instruments must operate with this source of power. Vacuum tube amplifiers require 100 to 300 volts dc between the cathode and the anode; means of providing these voltages must be available.

A diode vacuum tube can function as a rectifier, or for some applications a semiconducting device like a selenium rectifier is preferred. In either case, the resulting direct current pulsates with the same frequency as the original ac. Figure 13-8 shows a half-wave vacuum tube rectifier and a full-wave rectifier which offers some improvement. Even full-wave rectification results in a pulsating dc voltage, however. A *filter* composed of inductances and capacitances or of resistances and capacitances is used to smooth out the fluctuations. The inductance or resistance tends to oppose any increase in the voltage, but the capacitors tend to oppose any decrease in voltage. The result is a direct current with some "ripples" but a current considerably smoother than the original pulsating dc.

If even better control is desired, an electronic regulating system can be used. Such a circuit employs gas-filled tubes, in which the voltage across the tube is independent of current, for reference. The voltage from the filter is compared continuously with the gas-filled tube, and any tendency to increase or decrease is counteracted. There are several means of achieving this control, some of which become quite elaborate.

In addition to the high voltage for the amplifier tubes, the power supply is likely to provide a set of low voltage ac or dc supplies for heaters in vacuum tubes or for other purposes.

AMPLIFIERS   The basic unit of the vacuum tube amplifier is the triode. The signal to be amplified is fed into the grid of the tube, where it influences the passage of electrons between the cathode and the anode. Signal fluctuations appear on the anode amplified 30 to 100 times. Several stages are frequently used in amplifiers, arranged so that the output of one stage is further amplified by the next stage. Total amplification may be 10,000 or more times.

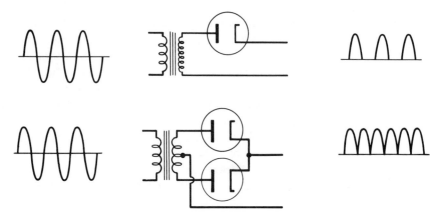

FIGURE 13-8   *Rectifier circuits. Top: half-wave, using one diode. Bottom: full-wave, using two diodes (commonly enclosed in the same glass shell).*

The easiest amplifiers to build are those for alternating signals. The successive stages are connected through capacitors so that the alternating signal can pass, but any direct current cannot. A direct-coupled amplifier is more useful in biology because it will amplify direct currents as well as alternating currents. It is unfortunate that a very similar abbreviation, D. C., is used for direct-coupled as for direct current. Direct-coupled amplifiers come with built-in problems, and their design is best left to the electronics engineer. The problems are so serious that once they are solved the result is a very fine amplifier indeed.

Biological amplifiers have some special requirements. The "hi-fi" amplifier is constructed so that it provides the same amplification regardless of frequency over a wide range. The "addict" is proud to state that his amplifier is "flat" from 5 to 50,000 cycles per second. Biological signals are direct current, or very low frequency alternations, so this fine frequency response is wasted. If a response like a nerve potential is to be measured, then somehow the nerve cell must be connected to the grid of the first amplifier stage. A small voltage (bias) must be applied to the grid, however, and this bias is likely to have "unbiological" consequences in the nerve cell. Or, to put somewhat the same idea in a different way, if a pair of electrodes is attached to a cell or tissue, and the impedance through the external circuit is lower than the impedance of the cells, then any electrical potential in the cells results in a current in the external circuit instead of in the cells where it belongs. Therefore, some form of high impedance input circuit must be used with biological systems in order to prevent the instrument from influencing the cells.

Transistor amplifiers offer several advantages in biological experimentation by partially overcoming some complex problems. One of the difficulties in the use of a conventional amplifier for measuring biological reactions is the set of electrical properties at the junctions between the electrodes and the cells. Transistor amplifiers permit the use of electrode-tissue junctions with less tendency to "damp" or obscure the biological signal. Another advantage is that transistor amplifiers give their best performance in the range of frequencies encountered in biological experiments. Finally, the small size and low power requirements permit the use of transistor amplifiers in situations where a vacuum-tube amplifier would be impractical. Transistor engineering is very complex, so even the biologist who is quite competent in electronics does not try to design his own circuits.

POTENTIOMETRIC TECHNIQUES A number of laboratory instruments measure electrical potentials (or changes in potentials) which result from chemical reactions. Consider, for example, a piece of copper wire with one end in a solution of a copper salt. Copper atoms exist in an equilibrium, $Cu \rightleftharpoons Cu^{++} + 2e^-$. Copper atoms will be deposited on the wire and removed from the wire at random. Once the equilibrium is established there will be no net increase or decrease in the amount of copper metal or Cu^{++}. Any increase in the concentration of Cu^{++} or e^- leads to a deposition of Cu metal on the wire. Now, consider a similar system containing another metal, say zinc, in a solution of its salt. The same kind of equilibrium will be established, $Zn \rightleftharpoons Zn^{++} + 2e^-$, but the proportion of metal and ions is different. If these two solutions are placed in the same container and the two wires are connected outside the solution, interesting things start happening. Zinc atoms are more likely than copper to exist as ions. If we compare the two equations above, one is more likely to run forward, the other backward. At the zinc wire, Zn^{++} ions depart into the solution leaving behind two electrons. At the copper wire,

Cu^{++} ions pick up electrons from the wire and become Cu metal. As long as the wires are connected, electrons flow through the external circuit from one electrode to the other. The pair of chemical reactions, then, has generated a voltage which produces a current in the external wire. Almost any metal would work as one member of such a pair. In fact it need not even be a metal because hydrogen gas and hydrogen ions reach the same kind of equilibrium. The hydrogen electrode is used as a reference point in potentiometric measurements. A number of kinds of biological reactions produce similar effects, directly or indirectly. These reactions can be followed by measuring the potentials developed. In practice one of the electrodes is commonly the calomel electrode $(Hg_2Cl_2 \cdot KCl)$, and the other is a platinum wire which can absorb electrons from an oxidation-reduction reaction occurring in the solution.

The potential developed in such a system is measured with a potentiometric circuit. The voltage to be measured is compared to some standard voltage by means of a bridge-like arrangement of resistors.

The electrical measurement of *pH* depends on a similar electrode reaction. A solution of known hydrogen ion concentration is contained in a tube of glass (the glass electrode) in contact with a silver-silver chloride couple. A calomel electrode is used as the other "half-cell." If the glass electrode is placed in a solution containing hydrogen ions, as all aqueous solutions do, hydrogen ions move through the glass membrane. They move inward or outward, depending on the *pH* of the solution; the result is a potential which can be measured. Within the useful range, the potential developed is a remarkably consistent function of the *pH* of the solution.

An important use of potentiometric techniques is to follow oxidation-reduction reactions. Oxygen may or may not be involved.

CIRCUIT DIAGRAMS   Most electronic instruments are accompanied by instruction books containing the complete circuit diagrams. These diagrams increase the likelihood that the operator can maintain the instrument and make necessary repairs and adjustments. Symbols have become more-or-less standardized, which makes the task easier. Figure 13-9 is arranged to

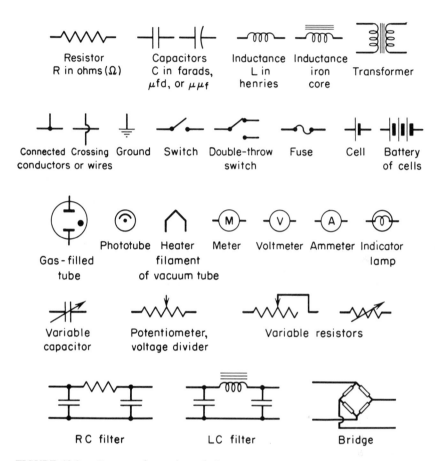

FIGURE 13-9   *Commonly used symbols.*

show some commonly used symbols as well as some representative circuit components. An interesting and very instructive exercise is to draw a circuit diagram by tracing the parts and connections of a finished electronic device, even a radio.

SELECTED Corning, John J., *Transistor Circuit Analysis and Design.*
REFERENCES Englewood Cliffs, N. J.: Prentice-Hall, Inc., 1965.

Lion, Kurt S., *Instrumentation in Scientific Research: Electrical Input Transducers.* New York: McGraw-Hill Book Company, Inc., 1959. A truly remarkable coverage of a difficult subject: comprehensive, detailed, factual, and yet readable.

Offner, Franklin F., *Electronics for Biologists.* New York: McGraw-Hill Book Company, Inc., 1967.

Phillips, Leon F., *Electronics for Experimenters in Chemistry, Physics, & Biology.* New York: John Wiley & Sons, Inc., 1966.

Suprynowicz, V. A., *Introduction to Electronics: for Students of Biology, Chemistry, and Medicine.* Reading, Mass.: Addison-Wesley Publishing Company, Inc., 1966.

Whitfield, I. C., *An Introduction to Electronics for Physiological Workers,* 2nd ed. London: Macmillan & Co. Ltd., 1960. Whitfield uses the British terminology (for example, he refers to vacuum tubes as "valves"), but some of his explanations are the most lucid to be found anywhere.

Whitfield, I. C., *Manual of Experimental Physiology.* New York: The Macmillan Company, 1964.

**OTHER
PHYSICAL
METHODS**

Two physical methods, unrelated to each other, are described in this chapter. Although each of the descriptions could easily be expanded to a whole chapter, for various reasons I have chosen to keep the accounts short.

The analysis of crystals by X-ray diffraction is a method not likely to be used directly by students in biology. Even the conceptualization of the relationship between structure and diffraction patterns is difficult, as is an understanding of the set of manipulations and computer operations that produces the final three-dimensional picture. Furthermore, the crystalline state is one in which the important biological molecules are not likely to occur in cells, and therefore the best picture we have is only indirectly related to the biologically active structures. Nonetheless, the contributions of X-ray diffraction studies to our understanding of the three-dimensional structure of enzymes and other important biological molecules is overwhelming. Thus at least a brief description seems appropriate, to help the student understand what he reads.

Electrophoresis is included here, not because the biologist is unlikely to use it, but to separate it from chromatography with which it is often confused. Electrophoretic separation depends on subtle differences in molecular properties. It has been possible to demonstrate in some instances that whole families of recognizably different molecules exist where no differences had been suspected after purification by less powerful techniques.

**X-RAY
CRYSTALLOGRAPHY**

The diffraction of X-rays is analogous to the diffraction of light rays. The paths of X-rays are changed by regions of high electron density, as in the vicinity of any moderately large atom. If atoms are arranged in a crystal in even repetitive spacings, the spaced atoms diffract X-rays in much the same way as a grating diffracts light. Recall from page 106 that $n \sin \alpha = \lambda/d$. Since the velocity of X-rays is about the same in any medium likely to be encountered, the refractive index, n, is about 1. The spacing, d, in a crystal and the wavelength, λ, of X-rays are of the same general order of length. Thus if X-rays are diffracted by the crystal, the angles, α, of diffraction are of a convenient magnitude to be measured.

205 Crystals are three-dimensional structures and diffraction patterns are complex. The ordered structure can be simplified for the purposes of illustration as shown in Figure 14-1. Here the dots represent atoms in one plane of a crystal. X-rays impinging on the crystal are diffracted along the paths shown. Even in two dimensions it is apparent that several different diffraction angles are possible. If the third dimension is considered, the pattern becomes even more complex. For this reason, the analysis of X-ray diffraction patterns is quite difficult. The effort is expended, however, because this is about the only available method for locating the individual atoms in the crystal.

The design of the X-ray camera is relatively simple in principle. An X-ray tube produces a beam of X-rays; metal foil filters isolate the chosen wavelengths. A sample crystal to be analyzed is placed in the holder diagrammed in Figure 14-2. A circular sheet of photographic film, with a hole in the center, records the diffracted X-rays. From the positions of exposed regions of the film, diffraction angles and intensities can be calculated. In studying a given crystal, a series of pictures would usually be made with the crystal held in different orientations.

If the crystal were rotated during the exposure, the film would show a pattern of concentric circles. The same effect is achieved if a powder made up of a large number of tiny crystals is photographed. The small crystals will be randomly situated in every possible orientation. In this case the X-ray picture is called a "powder diagram" and is used to test whether a given material is crystalline or amorphous. Since the ability to crystallize a protein is an important test of purity, such powder diagrams may be useful criteria of purity for biological chemists.

The formulation of a model for the three-dimensional structure of even a simple crystal like sodium chloride from the information available in X-ray diffraction pictures is no easy task. The atoms indicated in rows in Figure 14-1 are actually planes in three dimensions. Several simplifications must be made.

The task really becomes formidable if we examine crystals of complex substances like proteins or nucleic acids. The interatomic spacings still exist in regular patterns. A protein molecule contains so many atoms, however, that it seems hopeless ever to unravel the complex pictures. Furthermore, the electron clouds of C, H, O, and N atoms are not partic-

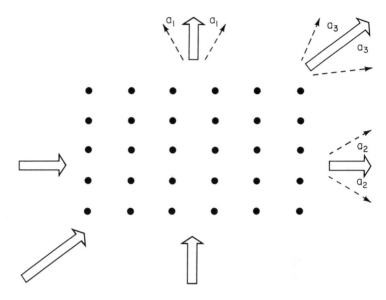

FIGURE 14-1   *Spacing and diffraction in one plane of a hypothetical crystal.*

ularly dense, compared to those of heavy metal atoms, so
that the diffraction is weak.

Despite the great difficulties, it was recognized many years
ago that crystalline proteins often occurred with helical
patterns in at least part of the molecule. Pauling formulated
the α-helical model for polypeptides in the late 1940's and
early 1950's. Without the careful X-ray studies of crystalline
nucleic acids, the formulation of the Watson–Crick double-
helix model for DNA would never have been possible.

Given the difficulties of technique and interpretation, then,
the formulation of models for the complete three-dimensional
structure of proteins is a remarkable achievement. Yet this
has now been accomplished. Controlled modification of the
protein and several other steps were necessary, followed by
computer analysis of the data. Kendrew has shown us a
complete model for myoglobin. Perutz describes the analytical
procedures used in a similar study of hemoglobin in a
readable paper listed at the end of the chapter.

ELECTROPHORESIS   If an electrical field is imposed across a liquid containing
charged particles, these particles will migrate in the liquid.
Negatively charged particles migrate toward the positive

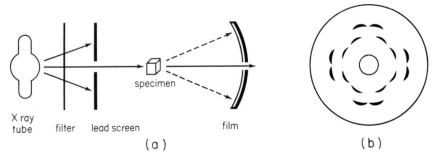

X ray
tube   filter   lead screen   film

(a)   (b)

FIGURE 14-2   *(a) Essential features of an X-ray diffraction apparatus. (b) A simplified diagram to show the general appearance of a film after processing. Actual crystals usually show many more spots.*

electrode; positively charged particles migrate toward the negative electrode. This behavior is the basis of an important method, known as electrophoresis, for separating materials. Proteins were the first and still the most important group of compounds to be studied.

Proteins are readily separated by electrophoresis because of the nature of the protein molecule. It is composed of a large number of amino acids, some of which possess side chains having an acidic —COOH group, others of which contain basic groups of one kind or another. The dissociation of these groups depends upon *pH*. At low *pH*, the carboxyl group will be associated (—COOH), and basic groups, such as —NH_2, will carry an additional proton (—NH_3^+). At high *pH*, the carboxyl groups dissociate (—COO^-), as do the basic groups. The net charge on the molecule thus depends strongly on *pH*. At some *pH* value, the isoelectric point, intermediate between the extremes, positive and negative charges balance each other.

The earliest electrophoretic studies were performed in glass tubes filled with a buffer solution with an electrode inserted in each end. A small amount of protein solution was injected through an opening approximately midway between the two ends of the tube. When the direct current field was applied, the protein molecules would migrate toward one pole or the other, depending on the sign of the net charge on the molecules. The rate of migration was followed by a complex optical system which detected the changes in refractive index associated with differences in the concentration of the protein solution. The rates were found to depend on the magnitude of the charge, the molecular weight, the shape of the particles, the viscosity of the solution, and several other physical factors.

Paper or gel electrophoresis is more commonly used today. Better resolution is available, because the components move in relatively discrete zones when the buffer is stabilized by the addition of the solid supporting phase. Mixing by convection is avoided.

PAPER
ELECTROPHORESIS

A strip of filter paper, moistened with buffer solution, is suspended between two reservoirs of buffer solution, with one end dipping into each solution. Electrodes are placed in the buffer reservoirs, usually physically separated from direct contact with the paper in order to avoid complications from possible electrode reactions. A small amount of the mixture to be separated is applied near the center of the paper, the voltage is applied, the buffer solution conducts a current, and the components move in one direction or the other, at rates determined by their physical properties.

Several kinds of paper electrophoresis chambers are available commercially. The direct current voltage, the wetness of the paper, and the amount of the mixture applied must all be given careful attention. Some difficulties which might be encountered include heating by virtue of the electric current, evaporation, and the electrical migration of the buffer solution. Various means have been devised to overcome these difficulties. If a commercial instrument is to be used, it is best to follow the recommendations of the manufacturer.

Continuous paper electrophoresis can be performed in an apparatus where a sheet of paper is hung vertically, as diagrammed in Figure 14-3. Buffer is dripped into the paper at the top to prevent drying. Electrodes are attached at the sides. The mixture is applied at the top in a very fine stream. As the components in the mixture migrate downward by gravity, the side-to-side electric field pulls them one way or the other, forming an angular path dependent on net charge and other properties. Separated components drip into test tubes from the points of the bottom of the paper.

Paper electrophoresis has been used in conjunction with chromatography in the technique known as "fingerprinting."

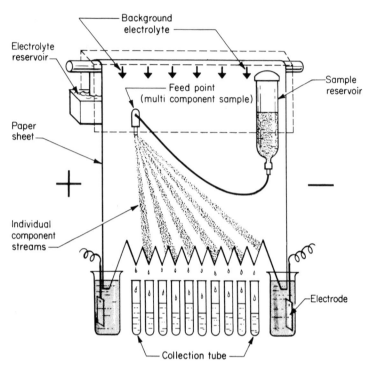

FIGURE 14-3   *"Curtain" electrophoresis for continuous separations. In operation the paper sheet is enclosed in a case. The paths taken by several components are shown; those most strongly positively charged move farthest to the right.*

A mixture is applied to a corner of a large sheet of paper and subjected to electrophoretic separation in one dimension, chromatographic separation in the other. When the amino acids resulting from the hydrolysis of a protein are separated by this technique, each amino acid comes to rest at a characteristic position. Geneticists use fingerprinting to learn about the differences in the amino acid composition of proteins which have been altered by mutation.

GEL
ELECTROPHORESIS   The apparatus for electrophoresis on starch gels is similar in many respects to that for paper electrophoresis. Specially-prepared starch is cast in slabs impregnated with buffer. Buffer solution is carried to the ends of the slab, and electrical

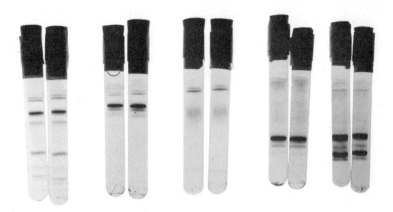

FIGURE 14-4   *Stained discs of proteins on polyacrylamide gel. The cylinders of gel are held under liquid in the test tubes to prevent their drying out. These are the proteins in samples of saliva from different sources. (Courtesy C. Bonilla and R. Stringham)*

connection is established, by means of paper wicks or some similar device.

Electrophoresis on starch gels, and on paper as well, can be adapted to a number of classes of compounds other than proteins. Amino acids, organic acids, some carbohydrates, etc. might be accommodated by the technique. If the molecules are not naturally charged, it is often possible to modify them chemically, or to form complexes which are charged.

Electrophoresis on polyacrylamide gels yields very high resolution of proteins. The polyacrylamide is polymerized inside small glass tubes. The tubes are then arranged upright, with the bottom end in one reservoir of buffer, the top connected to another reservoir. Small samples of proteins are separated electrophoretically, and stained to determine their positions. In the cylinders of gel, the protein zones appear as a stack of discs (Figure 14-4); this method is sometimes called disc electrophoresis.

This technique is frequently employed in clinical laboratories to analyze blood proteins for diagnostic purposes, as well as in the research laboratory.

PROBLEM   1. In disc electrophoresis on polyacrylamide gel, the separated proteins are commonly located by staining with a dark dye. A number of proteins are highly colored because a "chromophoric" group is attached as part of the molecule. Hemoglobin is one example, but there are many others to be found in various animals and plants. If you have the requisite equipment, try running some of these proteins through the gel, to see if they can be located by virtue of their own color. This could allow you to observe a separation as it is occurring, and at the same time, to avoid any possible side effects of the staining procedure.

SELECTED   Nuffield, E. W., *X-ray Diffraction Methods*. New York:
REFERENCES   Interscience (Wiley), 1966.

Perutz, M. F., *The Hemoglobin Molecule*. Scientific American Offprint No. 196 (Scientific American, November, 1964).

Setlow, R and E. C. Pollard, *Molecular Biophysics*. Reading, Mass.: Addison-Wesley Publishing Company, Inc., 1962. Has a good chapter on X-ray analysis.

Nerenberg, S. T., *Electrophoresis: A Practical Laboratory Manual*. Philadelphia: F. A. Davis Company, 1966.

Experimental data are meaningful only if they can be compared to some standard of reference and if general interpretations can be drawn from them. In all the sciences it has become customary to express data in certain sets of standard terms, just as the chemist expresses an amount of gas as the volume it would occupy at $0°$ C and 760 mm Hg. Biologists, whose materials are nonstandard, naturally have greater difficulty standardizing their figures. A chemist can speak of concentrations of solutions in moles per liter, but what is the concentration in moles per liter of the potassium ions in a single cell? Or what is the concentration in moles per liter of the *Chlorella* cells in a Warburg manometer flask?

Biologists must carefully specify the exact conditions under which a measurement was made if there is to be any hope of repeating the measurement. This can be accomplished by carefully detailed descriptions, but it can be done somewhat more simply by a careful choice of units in which to express results.

AMOUNTS OF
BIOLOGICAL
MATERIAL

If a 20-Kg dog eats 1 Kg of food a day, two 20-Kg dogs should eat about 2 Kg. In many instances such relationships hold reasonably well. Most of the computations expressed in this section are based on the assumption that the magnitude of an effect is directly related to the amount of biological material. Unfortunately, a 40-Kg dog is not likely to eat the same amount as the two smaller dogs for a variety of reasons. The assumption made above must be used with considerable care.

A direct comparison of two experiments is possible only if the same amount of the same kind of living material was used in both cases. An indirect comparison can be made by recalculating both sets of results in terms of some standard amount of material. Simply weighing the cells in both experiments makes such a comparison possible since the results can be expressed in terms of amount of change per gram of cells. Cells or tissues weighed in the living condition yield a "fresh weight."

Suppose we were measuring the amount of a certain ion absorbed by slices of potato tuber, and we wished to compare two batches of potatoes from different sources. The absorption of ions could be expressed as a number of grams (or milligrams) per gram (fresh weight) of tissue. Because one

213

batch of potatoes might contain a much larger amount of water, relatively, than the other, a direct comparison on the basis of fresh weight could be misleading. A more realistic measure of the amount of potato tissue, in this case, is the dry weight obtained by drying the tissue in an oven after the experiment is completed. Alternatively, one sample of each batch of potatoes could be used in the experiment and another sample of each, equal in fresh weight and volume, could be dried. Even this method is not entirely adequate. The absorption of ions sometimes depends upon metabolic activity, which in turn depends upon the relative concentration of enzymes present. If one kind of potatoes has a large dry weight, but most of this weight is metabolically-inactive starch, the results expressed on the dry weight basis are not very useful. A better comparison would be based on the amount of protein nitrogen per unit of potato tissue. The amount of protein would indicate the amount of enzymes present; thus, results expressed as milligrams of ions absorbed per unit of nitrogen are more realistic. This example illustrates a rather common dilemma in biological experiments. As the experimental results become more directly comparable, it becomes necessary to make two sets of measurements in each experiment: measurements of the amount of living material, as well as the measurement of the phenomenon being studied.

MANIPULATIONS OF RAW DATA   The purpose of an experiment is to answer a hypothetical question, but the results are just a set of numbers. The question cannot be answered unless the numbers are in a form directly related to the form in which the question was stated. Several kinds of manipulations of the raw data are sometimes necessary. One type is the transformation of dimensional units described in the preceding section. Another type is a conversion of the data produced by an instrument into other terms. For example, a stripchart recorder might record millivolts when we really want to know a change in chemical composition. In the properly designed instrument, the actual output will be related to the information being sought by some unvarying mathematical relationship or "transfer function." In the ideal instrument, the output is directly proportional to the response, the transfer function is linear, and the only correction required is a multiplication by a constant.

We must put up with random errors or variations of measurement. We can minimize such errors or estimate the size of the variations, but we cannot eliminate this source of error and we cannot correct for it. A systematic error, however, leads to an inaccuracy of measurement which results from some defect in the standards used for comparison. The too-short meter stick was used as an example in Chapter 4. We can estimate the magnitude of errors of this sort and then correct for them. If we know, for example, how much too short our meter stick is, a simple computation gives us corrected values.

The following extensive example illustrates several kinds of manipulations of data. A set of measurements of rate of photosynthesis was made using the manometric method (Chapter 10). The data were recorded on a printed form (one suggested by Umbreit, Burris, and Stauffer *), and some of the computations were performed on the same sheet. Figure 15-1 shows such a record sheet in an abbreviated form. The thermobarometer vessel contained only water. The other manometer contained *Chlorella* cells suspended in $KHCO_3$ solution so that the only gas exchange affecting the pressure was the production of oxygen. The constant, $K_{0_2} = 1.32 \ \mu l \ O_2/mm$ of manometer fluid, is a transfer function. The "raw data" include the time, the pressure on the thermobarometer, and the pressure on the experimental manometer. The second column under the thermobarometer and also under the experimental manometer is the amount of change since the beginning of the experiment. In the third column under the experimental manometer the first correction is made. The pressure change exhibited by the thermo-barometer represents a systematic error, that is, a change in room air pressure since the beginning of the measurement. Application of this correction changes the experimental values slightly. The numbers in the last column under the experimental manometer were derived by multiplying by the constant, an operation which converts (corrected)

* W. W. Umbreit, R. H. Burris, and J. F. Stauffer, *Manometric Techniques,* 4th ed. (Minneapolis: Burgess Publishing Company, 1964).

	Thermobarometer (TB)		Experimental manometer			
Temp: 20 °C Light: 150 w. Refl. Flood at 10 cm.	3 ml. H₂O		3 ml. Chlorella suspension in KHCO₃ 11 mm³ cells per ml. 1.7 μg chlorophyll/mm³ cells			
			KO₂ = 1.32 μl O₂/mm.			
Time	Time since beginning	Manometer reading	Change since beginning	Manometer reading	Change since beginning	Corrected by TB	Multiplied by constant
10⁰⁰	0	150		150			
10⁰⁵	5	150	0	168	+18	+18	24
10¹⁰	10	152	+2	189	+39	+37	49
10¹⁵	15	153	+3	208	+58	+55	73
10²⁰	20	151	+1	225	+75	+74	98
10²⁵	25	151	+1	244	+94	+93	123
10³⁰	30	152	+2	263	+113	+111	146
						ave./5 min	24.3

FIGURE 15-1   *Record of a manometric measurement of photosynthesis; student data.*

millimeters of manometer fluid to microliters of oxygen at standard conditions.

Further computations can be made. We know the amount of oxygen produced in thirty minutes by this lot of *Chlorella* cells. The average rate of photosynthesis (in μl O_2/hr) is given by multiplication. We know the volume of the cells and the amount of chlorophyll they contain, so we can calculate the rate in any terms we choose.

In some experiments it is possible to combine all the corrections and computations in a single equation. Using the number of millimeters of manometer fluid (corrected by thermobarometer reading) from Fig. 15-1, we can compute the rate of photosynthesis in μl $O_2/\mu g$ chlorophyll $\times$ hr by the following equation:

$$\left(\frac{111 \text{ mm manometer fluid}}{30 \text{ min} \times 3 \text{ ml cell suspension}}\right)$$

$$\left(\frac{1.32 \ \mu l \ O_2}{1 \ mm \ manometer \ fluid}\right)\left(\frac{60 \ min}{1 \ hr}\right)$$

$$\left(\frac{1 \ ml \ cell \ suspension}{11 \ mm^3 \ cells}\right)\left(\frac{1 \ mm^3 \ cells}{1.7 \ \mu g \ chlorophyll}\right)$$

$$= 5.2 \ \frac{\mu l O_2}{hr \times \mu g \ chlorophyll}$$

Some of the dimensions and some of the numbers cancel, giving us the numerical rate in the desired terms. Arranging all corrections and conversions in this manner provides a dimensional check. If we performed each conversion separately, we might end with a rate in terms of oxygen production per square milligram of cells, or another such ridiculous unit.

AIDS IN CALCULATION   Computations are likely to be long and involved in any extensive experiment. If an aid to calculation can be used without sacrifice of accuracy, it certainly should be employed. The figures resulting from biological experiments are subject to natural variation, and there is no need to maintain precision to twelve significant figures. Since three significant figures are usually adequate, slide rule accuracy is good enough.

The biologist is likely to use his slide rule for multiplication and division and for finding squares, square roots, and logarithms. The trigonometric scales probably are used only rarely. Even this limited use, however, makes the slide rule a valuable instrument.

For problems requiring more significant figures, or for sequences involving additions and subtractions which cannot be done on a slide rule, the electrical calculators are a great help. These machines can add, subtract, multiply, and divide automatically. With more or less ease, square roots can be determined. A competent operator can make calculations very rapidly with few mistakes. Electrical calculators are used extensively in statistical computations.

MATHEMATICAL TREATMENTS   Physical theories are most desirably expressed as simple mathematical equations. Biologists would like to present their theories in the same way, but so far nothing has appeared in biology with the beautiful simplicity and profound generality of $E = mc^2$. Nonetheless, on a lower level, experimental results and theoretical interpretations of the data can fit simple equations quite well.

Analytic geometry presents a variety of equations for geometrical figures. For example, if a graph is plotted on rectangular coordinate paper, the values of y on the vertical axis bear some natural relationship to the values of x on the horizontal axis. If $y = mx + b$, the graph is a straight line, m is the constant slope, and intercept b is the value of y when x is zero. The slope can be positive or negative.

The straight line or linear relationship is very common in the laboratory. A verbal expression indicating the same relationship is "y is directly proportional to x." The data in Fig. 15-1 can be used as an example. At zero time, the manometers contained some amount, b, of oxygen, although we were not concerned about this amount. After some time, it was apparent that the amount of oxygen produced in each five-minute interval was about the same. The slope, then, is the average change per five minutes, or 24.3 μl O_2/5 min. The total amount of oxygen in the vessel at the end of the measurement is

$$\underbrace{\frac{24.3 \ \mu l \ O_2}{5 \ \text{min interval}}}_{m} \times \underbrace{6 \ (5 \ \text{min intervals})}_{x}$$

$$+ b \ \mu l \ O_2 \ \text{at start} = 146 + b \ \mu l \ O_2$$

$$+ b \qquad\qquad\qquad = y$$

In our previous calculations, we assumed that b was unimportant and measured only the change since the beginning. This assumption does not change the equation; it merely assigns a value of zero to b, whence $y = mx + 0$.

Sometimes the actual expressions for which x, y, m, and b stand are exceedingly complex. A part of the genius of the theoretically-minded biologist lies in the ability to recognize simple equations in these complex expressions or to convert some more complex relationship into a straight line.

The straight line is probably the most common relationship, partly because many experiments are set up to test for such a relationship. Perhaps the next most common equation is that for the hyperbola. The basic equation for the hyperbola

$$\frac{x^2}{a^2} - \frac{y^2}{b^2} = 1$$

gives a curve like that in Fig. 15-2. Rotation of the axes yields another expression, $xy = k$, which is graphically represented in Fig. 15-3. This expression is quite commonly represented in the results of experiments because it indicates

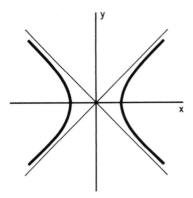

FIGURE 15-2 $\dfrac{x^2}{a^2} - \dfrac{y^2}{b^2} = 1$

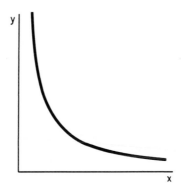

FIGURE 15-3 $x\,y = k$

that y is inversely proportional to x, or $y = k/x$. Still another transformation involving rotation and translation of the axes yields the curve in Fig. 15-4, with the equation

$$y = \frac{H\,x}{K + x}$$

where H and K are constants which establish the values at which the two arms of the curve become asymptotic. When $x = 0$, $y = 0$. When x is large, y is very nearly equal to H. Such a curve might result in an experiment where, for example, an over-all rate is controlled and limited by two separate factors. When x is small, the rate is limited by x,

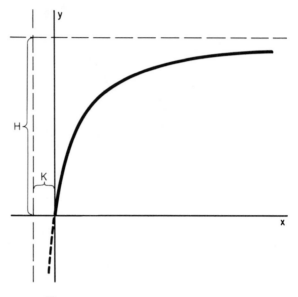

FIGURE 15-4 $y = \dfrac{Hx}{K + x}$

so that the rate increases rapidly as x increases. When x is large, some other factor limits the rate, so further increases in x make little difference in the rate.

Exponential and logarithmic expressions also occur frequently in the treatment of experimental results. The general exponential equation, $x = a^y$ (where commonly $a = 2$, e, or 10) can be written in the form $y = \log_a x$. This equation produces the curve of Fig. 15-5. Because of its curvature,

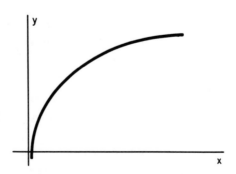

FIGURE 15-5 $y = \log_a x$

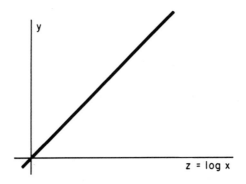

FIGURE 15-6   *y = z, where z = log$_a$ x*

this graph is difficult to interpret. Notice the similarity to Fig. 15-4. The logarithmic curve is easily converted to a straight line however. If we set $z = \log_a x$, then $y = z$, which is an equation for a straight line (Fig. 15-6). If it is suspected that the data fit a logarithmic curve, plotting log x against y tests this idea by either producing a straight line or failing to do so. This graph is easiest to construct on semilog paper, a special graph paper on which the subdivisions on one axis are logarithmic rather than equally spaced. Log-log paper, in which both axes have logarithmic scales, is sometimes useful; it is almost a laboratory joke, however, that log-log paper will make a straight line out of any set of data. The logarithmic relationship could, of course, exist in the opposite form, log $y = x$.

Differential equations handle all the considerations above a good deal more rapidly than algebraic treatment. Most experimental biologists eventually take courses in differential and integral calculus so that they can take advantage of this useful mathematical tool.

<table>
<tr>
<td>

16

**STATISTICAL
TREATMENTS**

</td>
<td>

A formal mathematical analysis of the results of experimental research is almost a necessity if the data are to be interpreted in terms of some hypothesis. This analysis is needed because of the natural variation in the numbers obtained from all measurements. Most hypotheses cannot be answered with an unquestionable "yes" or "no"; the variability of the results makes us say "probably." The formal system known

</td>
</tr>
</table>

as statistics, or more specifically as biometry, allows us to say "probably" with some estimate of how closely this "probably" approaches a definite "yes" or "no."

Biometry is now a method for the analysis of the quantitative data resulting from biological research. Although it is a branch of the division of mathematics known as statistics, biometry includes and uses almost all of statistics. "Statistics" originally were numbers, particularly numbers applied to populations of people. Thus "vital statistics" include such figures as the number of births, deaths, marriages, the incidence of diseases, and the number of persons employed by the automobile industry. Modern statistics includes a great deal more than measurements or records concerning people. Even the grammatical usage of the word has changed. When "statistics" is construed as a plural noun, each statistic is a number or bit of information. "Statistics" used as a singular noun, however, refers to a field of mathematics. Usually we can judge the meaning from the context.

In addition to their usefulness in analyzing data, biometric principles are useful as a guide in planning experiments. If we know in advance what form of statistical analysis will be required, we can be sure to take enough measurements to assure meaningful results, without taking a wastefully large number of measurements. More important, perhaps, we can be sure to take the right kind of measurements.

PROBABILITY   Statistics, a field based upon the laws of probability, treats events that occur at random. Most people have an intuitive notion about simple probability and agree readily that an honest coin comes up heads half the time and that seven is most likely to appear in a roll of a pair of dice. Many people find it harder to believe, however, that a long string of heads does not increase the probability that the next coin will be

222   a tail. And yet this is one of the fundamental laws of prob-

ability. The professional gambler is able to stay in business because he knows this.

Truly random events occur according to probabilities that are inherent in the events themselves and are not influenced by outside conditions or time sequences. Any individual atom of a radioactive isotope has a certain probability of decaying within the next second. This likelihood is not influenced by the presence of other atoms of the same kind. Random events are completely unpredictable on an individual basis. If measurement is influenced only by random variations, it is just as likely that the measurement will be a little too large as a little too small.

In theory, events occur at random because they are not influenced by outside conditions. In practice, it is too easy for events to be affected by bias of some sort, and therefore not to occur at random. In the analysis of experimental results, procedures are used which assume that the errors occur at random. The error or variability of measurement can be treated statistically *only* if the variability is random. For this reason, special steps must be taken in planning the experiment to assure the randomness of the errors. Laboratory biologists do not commonly carry out formal randomization steps, but, instead, hope to obtain experimental results that answer the hypothesis even without statistical treatment of the data. Unfortunately, too many of them perform statistical analyses, even without prior planning. If the assumption of true randomness is ignored, the statistical analysis is not only meaningless but deceptive as well.

Statistics ideally treats populations of things or events. By a population we mean all the possible things in a particular class, just as the human population includes all the people. There would be an infinite number of possible repetitions of a certain measurement, and these together would constitute a population. A population has certain characteristics, including variability, that can be described in detail. Obviously experimental work with populations is impractical; usually we work with samples of populations. A sample of the human population might include all the people in a city or all the people in a room. The smallest sample is one person. The features of a sample are similar to the features of a population but obviously cannot be identical. If a sample is large, it more truly represents the population. Statistical procedures have been developed for dealing with populations and with large and small samples.

THE NORMAL CURVE One of the characteristics of a population, or even of a sample, is a natural variability. In general, the individual values in the population of numbers tend to clump around a certain average value, but some separate values will be very much larger or very much smaller. There are several ways of describing patterns of variation, but when dealing with populations, all these methods approach the bell-shaped "normal curve." A population of numbers is likely to be normally distributed, that is, to vary according to the normal curve.

Several "parameters" can be used to describe this normal distribution within populations, but two of these are most important. The value around which the other values seem to be centered is the average or mean. When speaking of populations it is given the symbol m. The mean is found by adding all the values and dividing by the number (N) of values. This process is expressed mathematically, by calling each value or variate $x_1, x_2, x_3, \ldots, x_N$. The symbol x_i stands for any value of x: x_1, then x_2, then x_3, and so on. The capital sigma (Σ) is used to indicate summation, so $\sum_1^N x_i$ means "the sum of all the x values from x_1 to x_N." The arithmetic mean then is

$$m = \frac{\sum_1^N x_i}{N}$$

In the normal curve, the total of all deviations above the mean is equal to the total of all deviations below the mean. If there is likely to be no question about the meaning, the simpler expression, $m = \Sigma\, x/N$, can be used in place of the one above.

Another characteristic of the normal curve is an amount of variability. Figure 16-1 illustrates two normal curves with the same mean but different variability.* The deviations from

---

* These two curves are actually the same curve, representing the same equation. The scale has been changed on one for the sake of illustration. The normal curve fits populations of data. Sample data only approach the normal curve—some samples have more variability than others.

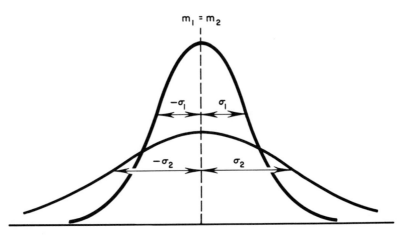

FIGURE 16-1   *A pair of normal curves having the same mean (m) but different standard deviations (σ).*

the mean are much greater and more numerous in the lower curve. This relationship is made quantitative by calculating the variance. Each variate differs from the mean by an amount $x - m$, and those values lying below the mean are negative. The negative signs are eliminated by squaring each deviation. The sum of the squared deviations, divided by the number of such values, yields the variance (σ^2).

$$\sigma^2 = \frac{\Sigma\,(x - m)^2}{N}$$

The square root of the variance is called the standard deviation σ.

The mean and the variance provide enough information to determine the shape of the normal curve.

PARAMETERS OF SAMPLES   Because the mean m and the variance σ^2 refer to the normal distribution of the population, they are rarely known precisely. They can be estimated from the characteristics of a sample or a series of samples. The better the sampling procedure, the better the estimate of m and σ^2. Each sample, even if it includes only one variate, has a sample mean, which might be nearly the same as the population mean or quite different. If a large number of samples is taken, the average of all the sample means provides a good estimate of the population mean.

The sample mean is calculated in the same way as the population mean but is given the symbol $\bar{x}$.

$$\bar{x} = \frac{\Sigma\, x}{n}$$

where n is the number in the sample.

If n is large, the variance of the sample (s^2) can be calculated in the same way as the population variance. The sample variance only approximates the variance of the population, however, and tends to underestimate the population variance by an amount equal to $(n-1)/n$. For this reason, sample variance is calculated by

$$s^2 = \frac{\Sigma\,(x - \bar{x})^2}{n - 1}$$

The standard deviation (s) of the sample is the square root of this value. A working formula which makes computation on an electric calculator easier is

$$s^2 = \frac{\Sigma\, x^2 - \dfrac{(\Sigma\, x)^2}{n}}{n - 1}$$

The standard error of the mean is a commonly used figure. It is defined as the standard deviation of a distribution of means. If a large number of samples is taken from a population, the means of the samples are distributed in a normal curve. The variance of the distribution of means, $\sigma_{\bar{x}}^2$, can be shown to equal σ^2/n. The standard error of the mean, then, is $\sigma/\sqrt{n}$. There are other ways of arriving at a value for the standard error; some of these will be discussed later with some other special statistics.

TESTS OF SIGNIFICANCE   The aim of statistical analysis is an estimate of the significance or meaning of the data. The analysis reveals the probability that the observed effects result from the experimental treatment, as opposed to pure chance.

Let us imagine an experiment as an example. We have observed that *in vitro* a chemical compound C combines with one of the enzymes involved in cellular respiration, but C is not a normal metabolic material. We can form the hypothesis that if C is introduced to living cells it should inhibit the normal respiration. The experiment involves a comparison of the activity of cells in the presence and in the absence of C. Now, the normal cells will vary, and one sample of cells might be quite different from another sample even if all the untreated cells are part of the same population. Suppose that the cells treated with C respire somewhat more slowly than the average of normal cells. The difference might result from the treatment, but it is also possible that the difference is the result of chance, that is, that even a sample of untreated cells could give a rate this much lower than average. A statistical analysis reveals the likelihood or probability that a difference this large or larger could result from chance.

If the mean and variance of the population are known, the test of significance becomes relatively easy. The area under the normal curve, or any portion of the total area, can be calculated. If we calculate the area included between a point 1σ below the mean and the point 1σ above the mean, about 68 per cent of the area under the curve is included. This means that 68 per cent of the variates lie between these limits (see Fig. 16-2). Between -2σ and $+2\sigma$, about 95 per cent of the cases are included, and 99^+ per cent lie between $\pm 3\sigma$. Other calculations can be made as well. For example, we could find a pair of lines, one on each side of the mean, that would include 50 per cent of the cases.

Our experiment provided us with a figure for the rate of respiration in cells treated with compound C. If we know the mean and variance of the respiratory rates in normal cells, we can calculate the probability that the observed results are different. In theory, it is impossible to know the mean and variance of the population but, if we have made measurements on a large number of samples, and assuming that σ is about equal to s, we can have a very good estimate of the parameters of the population. If the rate observed in treated cells is 4.5σ below the mean for untreated cells, then the probability that this sample belongs to the population of untreated cells is very small indeed. If the rate is only 1σ below the population mean, it would be dangerous to conclude that the treatment has had an effect because 32 per cent (100 per cent — 68 per cent) of the samples of the untreated population will deviate from the mean this much.

NORMAL DEVIATE   In the previous few paragraphs we have been speaking, without defining it, about a statistic known as the normal deviate. A value can be tested for significant deviation from the population mean by expressing the deviation from the

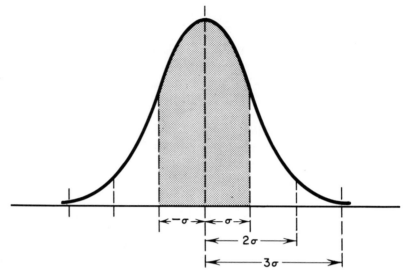

FIGURE 16-2   *The normal curve, showing the relationship between the standard deviation and the included area under the curve.*

mean in terms of standard deviations. The normal deviate (*c*) is

$$c = \frac{(x - m)}{\sigma}$$

which indicates that the value differs from the mean by a certain number of standard deviations. Published tables of probability tell us how frequently such deviates occur in the normal curve. In 5 per cent of the samples, *c* is 1.96 or greater; in 1 per cent of the samples, *c* is 2.58 or greater; and in 0.1 per cent of the samples, *c* is 3.29 or greater.

The normal deviate is also used to test the significance of the difference between the mean of a sample and the mean of a population. In this case, the standard error of the mean is used instead of the population standard deviation:

$$c = \frac{\bar{x} - m}{\sigma_{\bar{x}}} = \frac{\bar{x} - m}{\sigma/\sqrt{n}}$$

CONFIDENCE   In a statistical test of significance we select some arbitrary
LIMITS   probability limits. If a deviation as large as that observed occurs by chance in only 5 per cent of the cases, it may be safe to conclude that the deviation is the result of the experi-

mental treatment. In another experiment we might choose the 1 per cent level (probability $= 0.01$) of significance. The level of statistical significance chosen tells the confidence with which we can draw conclusions.

We can never say absolutely "yes" or "no," but only "probably" or "very probably." The actual choice of confidence limits depends upon the seriousness of drawing the wrong conclusion. It is possible to accept a false hypothesis or to reject a true hypothesis on the basis of chance variations alone. The rejection of a true hypothesis is called an error of the first kind, and the acceptance of a false hypothesis is called an error of the second kind. Depending upon the experiment, one error is more serious than the other.

As an example suppose we test the hypothesis, "Drug A is harmless to human beings." The drug will be released for public use only after statistical tests. If we select the $P = 0.05$ level of significance, we are 95 per cent certain that the drug is harmless. This is not sufficient confidence because the 5 per cent chance that the drug is actually harmful is too great a risk.

STUDENT'S t TEST   When the variance of the population is unknown and cannot be reasonably estimated, as is true in experiments involving small samples, the normal deviate test of significance cannot be used. An English statistician who signed his name "Student" derived a statistical test useful in such cases. A value of t is computed in a manner similar to the computation of c:

$$c = \frac{\overline{x} - m}{\sigma/\sqrt{n}} \qquad t = \frac{\overline{x} - m}{s/\sqrt{n}}$$

Here s is the standard deviation of the sample of which $\overline{x}$ is the mean. If the sample is large, s is very nearly equal to σ. If the sample is small, s is likely to be somewhat smaller than σ. The probabilities of t values arrived at in this way are thus slightly different from the distribution of c values. Tables of probability with which various values of t occur are used in this test of significance. The actual probability of a certain value of t depends upon the size of the sample; the test is said to have a certain number of degrees of freedom, equal to $n - 1$, where n is the sample size.

To test whether a certain sample belongs to a population, we calculate the value of t, look up the probability in the table, and then draw conclusions in the same manner as in the normal deviate test.

THE CHI-SQUARE (χ^2) TEST   In some kinds of experiments the data occur as discrete numbers rather than as continuously variable quantities. A genetics experiment might yield flowers that are either red or white, and the observation of the results consists of counting the two types. According to hypothesis, the two types should occur with a certain probability. For example, we might expect three reds for each white; the probability for red is $\frac{3}{4}$, and that for white is $\frac{1}{4}$. The χ^2 test is an approximation of a much more involved test for deciding how well the observed counts fit the hypothesis.

A χ^2 value is calculated for each class of events. In the experiment we observe a certain number of white flowers (o), but from the hypothesis we can calculate the most probable number (c). The χ^2 value is the square of the deviation from the expected value divided by the expected value:

$$\chi^2 = \frac{(o-c)^2}{c}$$

A similar calculation is made for each of the classes. The final χ^2 value is the sum of these individual numbers

$$\chi^2 = \frac{(o-c)^2}{c_{(white)}} + \frac{(o-c)^2}{c_{(red)}}$$

In this test, only one degree of freedom exists because if a flower is not white, it has to be red. In another experiment, flowers might be red, white, or pink; here there are two degrees of freedom.

A table of χ^2 values is used in tests for significance. If the value of χ^2 is larger than could be expected ($P = 0.05$ or 0.01) for this number of degrees of freedom, then the observed data do not fit this hypothesis. In the example the expectation was $\frac{1}{4}$ white and $\frac{3}{4}$ red. If the χ^2 value suggests that this hypothesis is incorrect, we could test some other hypothesis, such as $\frac{1}{2}$ white and $\frac{1}{2}$ red.

ANALYSIS OF VARIANCE   In the ideal controlled experiment all factors are constant except the one being tested. The significance of this varying factor is easily tested by one of the foregoing statistics. In practice, however, the effect observed in an experiment is likely to be influenced by several factors. A truly efficient experiment tests the effect of several factors at the same time. Simple statistical tests become extremely difficult in this case, however. Each factor might act independently or might interact with one or more of the other factors.

Analysis of variance is a system of treating such experimental data. Since the actual analysis is intimately tied to the design of the experiment, a discussion of analysis of variance is given in the following chapter on experimental design.

REGRESSION AND CORRELATION One of the most important things we might wish to find from an experiment is the relationship between a pair of variable quantities. In any experiment, changes in x might produce proportional changes in y. The relationship is a straight line, but which straight line best fits the results? If the data are very good, the points on the graph deviate very little from the line that best describes the relationship, and the line, called the regression line, can be drawn by eye. If the results include a larger error, visual inspection might suggest any of several lines to fit the points, as is true with the data shown in Fig. 16-3. Obviously x has an effect on y, and $y = a + bx$,* but what is the true slope b and the intercept a?

Lines are commonly drawn through such variable data by the method of least squares. The sum of the squares of all the deviations from the line is made to be a minimum. In other words, that line is found such that the sum of the squared deviations is smallest. If $y = a + bx$, and if we know a and b, the line is determined. In the least squares formula

$$a = \frac{\Sigma y \, \Sigma x^2 - \Sigma x \, \Sigma y^2}{N \, \Sigma x^2 - (\Sigma x)^2}$$

$$b = \frac{N \, \Sigma xy - \Sigma x \, \Sigma y}{N \, \Sigma x^2 - (\Sigma x)^2}$$

where N is the number of points. These are the basic equations; variations may be used in special cases, as when the number of points is small.

The correlation coefficient (r) was devised to express

* Note that the symbols here are different from those given in Chapter 15. The expression $y = mx + b$ is the same same straight line. Here a is substituted for b and b for m.

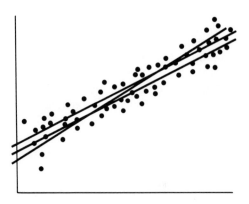

FIGURE 16-3   *A scatter diagram showing three alternative straight lines that might be drawn through the points.*

quantitatively the relationship between two sets of variables. It has values ranging from -1 to 0 to $+1$. If r is positive, larger values of x are also large with respect to y. Taller men are heavier, in general. Zero correlation means there is no consistent pattern between the x values and the y values. Negative values of r mean that values large in respect to x are small in respect to y.

The correlation coefficient actually measures the relationship of x and y in terms of their normal deviates (c),

$$r = \frac{\Sigma \, c_x c_y}{n}$$

This expression is calculated by means of the following approximation:

$$r = \frac{\Sigma (x - \bar{x}/s_x)(y - \bar{y}/s_y)}{n} = \frac{\Sigma (x - \bar{x})(y - \bar{y})}{n s_x s_y}$$

The computation is easier on the electric calculator if the equation is rearranged to

$$r = \frac{\Sigma xy/n - \bar{x}\bar{y}}{s_x s_y}$$

The correlation coefficient can provide valuable information but must be used with discretion. It is too easy to attribute a cause-effect relationship to a situation where r is large. For example, we might find a correlation between the amount of water flowing in a river and the number of automobiles sold. There is no conceivable way in which one of these could cause the other even if they might be related through a third factor, namely the season of the year.

OTHER STATISTICS   The mean is the most commonly used measure of central trend in experimental work. Two other measures are important in certain situations. The median divides the distribution of variates into two equal halves; half the values lie below the median, half above. The mode is that value which occurs most frequently. In the normal curve the three measures coincide.

Several methods other than the variance exist for estimating the variability of a distribution, or the "error." Experimental results are frequently presented in papers as (for example) 167 ± 13. The error figure (± 13) might have any of several meanings, and it can be interpreted by a reader only if the writer specifies how it was calculated.

The range, a useful estimate in some cases, is the difference between the largest value and the smallest. It gives quick information, and, if the ranges of two samples do not overlap, the difference between the samples is almost certainly significant.

Probable error is a deviation from the mean such that ± 1 PE includes one-half the cases. If the distribution of values is normal, ± 1 PE also includes half the area under the curve; from this, $PE = 0.6745\sigma$.

The standard error of the mean (SE) is the preferred statistic to be included in tables of data. If everyone agreed with this statement, then 167 ± 13 above would be the mean and the standard error of the mean. The standard error was calculated as $\sigma/\sqrt{n}$. Radioactive decay is almost perfectly random, and the number of counts determined is usually large. In this case a short-cut method of computing the standard error can be used. A certain number of counts (C) in a total time of t minutes yields an average rate of C/t counts per minute. The standard error is computed as $\sqrt{C}/t$. This value is so near to $\sigma/\sqrt{n}$ that the agreement between the two methods is used as a check on the randomness of the counting. This relationship holds because in the normal curve σ is almost equal to $\sqrt{m}$.

THE APPLICATION OF STATISTICS Because the use of statistical analysis depends upon a number of assumptions, the conclusions drawn can be no more valid than the assumptions. The assumption that the samples are random is a very important one but one too often forgotten.

Statistical signficance means only that the conclusion drawn is probably correct. Biometry is nothing more than a formal procedure for calculating probabilities. If the experimental data are good, that is, if the error or variability is small relative to the effect being measured, there is no need to analyze statistically. Statistical procedures are particularly valuable in those experiments where the errors are large or the effects are small.

Analysis of results and drawing conclusions require common sense. Blind acceptance of statistical significance can lead to ridiculous conclusions. I remember one-experiment in which three rows of bean plants were given identical experimental treatments, but one row produced more beans than the other two. The experimenter said, "I don't know what it means, but *it was significant at the 5 per cent level.*"

Most biologists are not sufficiently expert to know which test to use in every circumstance, and most cannot remember all the assumptions that went into the development of the various analytical procedures. It is a common practice to enlist the help of an expert statistician or biometrician. The time to do so, however, is before the experiments are performed. Most biometricians prefer to assist in the planning stages rather than being called in at the last minute in a desperate attempt to salvage something from a ruined experiment.

SELECTED REFERENCES Fisher, Ronald A., *Statistical Methods for Research Workers.* Edinburgh: Oliver and Boyd, 1950. Sir Ronald Fisher's name is at the head of every list of persons who have contributed most to biological and agricultural statistics and to experimental design. This book is a classic and still one of the best available for reference.

Freund, John E., Paul E. Livermore, and Irwin Miller, *Manual of Experimental Statistics.* Englewood Cliffs, N. J.: Prentice-Hall, Inc., 1960. This little book will be convenient for those people who understand statistics but cannot remember the details about equations and such.

Freund, John E., *Modern Elementary Statistics,* 3rd ed. Englewood Cliffs, N. J.: Prentice-Hall, Inc., 1967.

Woolf, C. M., *Principles of Biometry: Statistics for Biologists.* Princeton, N. J.: Van Nostrand Reinhold Company, Inc., 1968. Especially careful attention is paid to the assumptions involved in the derivations.

We have seen some of the statistical treatments whose primary purpose is to show the probability that interpretations are correct. This formal mathematical treatment gives a level of "significance" to the results. Experimental design is a term applied almost exclusively to the advanced planning of experiments in order to take advantage of the statistical procedures available. The statistical design takes into consideration all the assumptions that went into deriving the various statistics, thus making the significance tests valid. A more practical effect, perhaps, is that experimental design enables the research worker to gain more information with less effort. The designed experiment is as much more effective than the haphazard experiment as that same haphazard experiment is more effective than observation without experiment. The ideal experimental design, of course, is the one that yields the most information and the most reliable information with the least expense and effort.

The most extensive user of experimental design, and in fact the area responsible for the development of this branch of statistics, is agricultural research. A little reflection shows how necessary it is to the agricultural scientist that experiments be effective. A person interested in the effects of fertilizers on the growth of field crops can raise only one crop a year, and extensive test plots require large amounts of land. Imagine the serious result if badly planned experiments yield no information. Similarly, breeding and nutrition experiments with large animals are expensive and time-consuming even if carefully designed. Psychology and certain other areas related to human affairs also depend heavily upon experimental design, for somewhat the same reasons.

Laboratory biologists on the whole have been slow to adopt formal experimental design. One reason is that laboratory biologists customarily "design" their experiments without realizing it. It happens that experiments with microorganisms, small animals, and single plants can easily provide much information, so that very simple experimental designs are adequate. One is not likely to despair over the failure of an experiment to provide significant results if the experiment only cost an hour, one cake of yeast, and a dime's worth of chemicals. In most experimental laboratories, the experiments produce results faster than the experimenters can calculate what the data mean. The experimental biologist, however, is rapidly reaching the point where the remaining problems are the very difficult ones, and it may be necessary to use more complex experimental designs.

The following sections briefly describe some of the more commonly used experimental designs. The biologist is wise to seek the counsel of a statistician instead of attempting to design his own experiments. The statistician is likely to be able to foresee the failure of an experiment if inadequate replications or repetitions are performed, but he might also estimate that a proposed number of replications is more than enough and therefore wasteful.

SAMPLING AND RANDOMIZATION

It should be obvious by now that the organisms used and the measurements made in an experiment represent samples of much larger populations. The selection of individual animals or plants to be used in an experiment is thus an exercise in sampling. Regardless of the experimental design, very careful consideration must be given to the choice of samples because they must be random selections from the population.

It might seem a simple matter to pick a few rats from a cage full of rats and then to assign half of these to serve as controls and half to receive the experimental treatment. Unfortunately, it is virtually impossible to avoid personal bias. Maybe the rats selected are the ones that were easiest to catch. If half the rats are to be controls and half are to receive the experimental treatment, how does one decide which rats receive which treatment? In which group do you place the one rat that managed to bite your finger?

The purpose of randomization is to minimize the natural and unavoidable bias. In the selection of the rats it would be safest to let the flip of a coin decide whether any individual rat becomes a member of the experimental group or a member of the control group. Tables of random numbers prepared by computers, available in several of the statistics books, can also be used in randomization. Each individual in the large group must have an equal chance of being used in the experiment, and, of those used, each must have an equal chance of being a "control" or an "experimental." In many instances, it is advisable to employ a "double blind" procedure. The subjects are not only chosen at random, but are coded in such a way that the experimenter does not know which individuals are receiving which treatment until the results are recorded.

In many experiments the selection of the organisms is completely random. Other cases allow the use of special statistical techniques in which the deliberate selection of individuals and matching in pairs produces more useful information than the complete randomization. For example, suppose we needed five rats to serve as controls and five to receive an experimental treatment. We wish to test the effect of a new drug upon gain in weight by the rats. The amount of gain, however, might depend on the original weight of the rat. If we weigh the animals initially and then arrange the weights in a table in descending order, it is no difficult task to assign the rats in pairs that are more like each other than like all the rest. One member of each pair receives the drug, and we decide which one by flipping a coin. Each member of the pair must have an equal chance to receive the experimental treatment. If you find yourself wishing the coin would fall a certain way, you had better believe the coin rather than your feelings and wishes.

One of the beauties of the use of microorganisms or subcellular particulates is that the experiment is conducted with a very large sample. There is considerable variability among the individual cells in a yeast suspension, but the average behavior is surprisingly uniform. Only a few minor precautions are necessary in sampling from such favorable populations. For reasons that are difficult to understand, the first pipetteful is likely to be slightly different in composition from all succeeding pipettefuls. Perhaps the dry inner walls of the clean pipette have some influence. Because cells in suspension settle quite rapidly, swirling or stirring between samples is necessary. Obviously the number of cells per 5 ml is not uniform if the cell suspension in a 500-ml flask is more concentrated at the bottom than at the top. Finally, pipetting errors are likely either to increase or to decrease in taking a long series of samples. If all the experimental samples are taken first and all the controls later, bias may be introduced. The safe procedure, again, would be to withdraw a pipetteful of cell suspension, and then let a coin decide whether it is to be a control or an experimental sample.

Many people have been successful research biologists without ever paying much attention to the randomization of samples. The same people have had little need to analyze their data statistically. The reason, of course, is that the differences between populations (a population of controls and a population of experimentally-treated organisms) are very large compared to sampling errors and the variance of the populations. This difference is probably no accident because many of these experiments were very carefully thought out and steps were taken to make sure that if a difference existed it would be a large one.

SOME SIMPLE DESIGNS   The simplest of all experimental designs is the one most commonly used in the experimental laboratory, a simple comparison of two groups of measurements. The following experiment can be used as an example, and it can be seen that the same reasoning could be applied to many different experiments.

We wish to know whether maleic hydrazide has any effect on the respiration of yeast cells. We set up six manometers with a yeast suspension and then, taking care to randomize, add maleic hydrazide to one set of three and an equal concentration of a neutral salt to the other three. We measure the oxygen consumption at five minute intervals for thirty minutes. We now have samples from two populations. Each population has a mean (m_1 and m_2) and a variance (σ_1^2 and σ_2^2). The means obtained in the three manometers provide a good estimate of the population mean. The only estimate we have of the population variance is the variance of the sample s^2. The application of the t test tells us whether the difference between the means is significant and at what level of significance.

PAIRED DESIGN   A similar experiment could be performed as a class exercise. Imagine seven groups of students, each performing the experiments described above. For simplicity, let each group of students use only one pair of manometers instead of three. The data could be analyzed by lumping together all the figures for untreated cells and all those for cells treated with maleic hydrazide. Whether each group uses one pair of manometers or three, the variance is likely to be somewhat larger than before because of individual differences in technique, in reading manometers, and in timing the readings. It is possible that each group would find a significant difference, but the combined data would indicate no statistical significance. In this case it would be wise to match the results in pairs, with the experimental and control manometers of any one experiment compared to each other. The statistical treatment becomes slightly more complex because there is reason to expect some interaction between pairs of manometers. One group of students might pipette the yeast cells generously, for example, so all their results would be higher than average.

It is necessary to account for this covariance, as the interaction is called, but this can be done by a shortened method. For the sake of analysis we hypothesize that the two population means are equal, that maleic hydrazide has no effect. In the manometers of each of the seven groups there is a difference between manometer 1 (without maleic hydrazide) and manometer 2 (with maleic hydrazide). The seven differences make up a sample from a possible population of differences. According to the hypothesis, the mean of these differences is zero. Table 17-1 shows some numerical results which will make the analysis easier to follow.

TABLE 17-1   *results of a paired experiment*

group	μl O_2/30 min manometer 1	manometer 2	d	$d - \bar{x}_d$	$(d - \bar{x}_d)^2$
1	117	111	−6	0.9	0.81
2	112	107	−5	−0.1	0.01
3	111	107	−4	−1.1	1.21
4	106	100	−6	0.9	0.81
5	104	99	−5	−0.1	0.01
6	114	110	−4	−1.1	1.21
7	109	103	−6	0.9	0.81
$n = 7$	$\bar{x} = 110.4$	$\bar{x} = 105.3$	$\bar{x}_d = -5.1$		$\Sigma = 4.87$

The seven differences have a mean of 5.1 as opposed to a hypothetical mean difference of zero. We can apply the *t* test to find out if the observed difference could occur by chance. In order to compute *t*, we must find the standard error of the mean $(s/\sqrt{n})$. The standard deviation is calculated from the table as

$$s = \sqrt{\frac{\Sigma (d - \bar{x}_d)^2}{n - 1}} = 0.9$$

The difference between the two means $\bar{x}_1$ and $\bar{x}_2$ is the same as the mean difference $\bar{x}_d$, and

$$t = \frac{5.1}{s/\sqrt{n}} = 14.7$$

The number of degrees of freedom is $n - 1 = 6$, and in the t table we find that the chances of a t value this large are less than one in a thousand. Therefore, this particular difference between means does not belong with the distribution where the mean difference is zero. We must reject the hypothesis that the two means are equal, or in other words, we conclude that the differences observed are statistically significant and that maleic hydrazide does have an effect on the respiration of yeast cells. Incidentally, if the pairing design is not used, the difference between the two means is not significant.

MORE COMPLEX DESIGNS   The simple experimental designs are adequate when only one pair of conditions is to be tested. If a larger number of comparisons is to be made, however, performing one set of measurements at a time may become unduly laborious. With the application of a little ingenuity it becomes possible to make several comparisons in a single experiment and still have legitimate statistical tests available. In most of these instances, analysis of variance is used. In the following discussion, three general types of experimental designs with examples are described, and then a typical analysis of variance is performed on one of these.

COMPLETELY RANDOMIZED DESIGN   In this design, as the name suggests, the choice of samples and treatments is left entirely to chance. The following example will not only illustrate this design but can be modified later for some other designs. We are interested in studies using chloroplasts isolated from leaves. We have decided to use sugar beet plants and to raise the plants in a special constant-temperature chamber under artificial light. The hope is to provide uniformly active plants. We have our choice of soil, vermiculite (heat-expanded mica), or an aerated solution as a medium in which to raise the plants. We perform the following preliminary experiments to determine which medium provides the greatest weight of leaves.

V_3	L_4	S_3	S_1
V_1	L_2	V_4	L_1
V_2	S_2	L_3	S_4

FIGURE 17-1   *A completely randomized design. V, L, S stand for separate treatments; the locations were chosen by chance.*

Four plants (or groups of plants) are to receive each treatment. Since the space in the plant growth room is essentially uniform, the groups of plants are placed at random within the space available. They might be arranged on the bench in a pattern consisting of 4 rows of 3. Randomization could be achieved with *one* of a pair of dice in the following manner. Assign the numbers 1 to 4 to the plants in each of the three groups. Now, number the possible locations on the bench from 1 to 12. Starting with position 1, throw the die; use this throw to decide which of the three groups is to be placed here (1 or 4 for soil, 2 or 5 for vermiculite, 3 or 6 for liquid, for example). With a second throw, decide which plant of the four is to be placed here. Proceed to the second available spot and repeat the process. Continue until all twelve plants are assigned. If by chance all the plants in soil should be together, trust the die because dice are better judges of randomness than you are. The plants in Fig. 17-1 have been labeled S, V, and L for soil, vermiculite, and liquid, respectively. They were placed by exactly the method described. There are simpler ways of randomizing, but this serves as an example.

Now we allow the plants to grow until they have produced large leaves and then cut off the leaves and weigh them. The statistical analysis tells us whether there are any significant differences among the three treatments.

RANDOMIZED BLOCK   Use the same example as above. We may suspect that one end of the bench is very slightly warmer than the other end or that light intensity and humidity are not quite uniform.

V_1	V_3	L_3	S_4
S_3	L_1	V_2	L_2
L_4	S_1	S_2	V_4

FIGURE 17-2 *A randomized block design. Each vertical block includes each treatment once. Within the block the treatments are random. The blocks could also be arranged horizontally.*

S	V	L
V	L	S
L	S	V

FIGURE 17-3 *A Latin square. Any treatment is present in each row and in each column.*

In order to distribute the environmental variation more evenly among the plants, we divide the area into four "blocks" of three spaces each. One plant from each treatment goes into each block, but within the block the plants are randomized. One such finished design is shown in Fig. 17-2.

LATIN SQUARE

This is a modified randomized block which helps to overcome environmental variation in two directions. Suppose that temperature and light intensity vary in the plant growth room, not only from end to end, but from side to side as well. In this instance we arrange the space in rows and columns, making sure that each treatment occurs in each row *and* in each column. The Latin square need not be square, but, for simplicity, let us reduce the number of plants in each treatment to three. We then have three replications of each of three treatments, which can be arranged in a Latin square as shown in Fig. 17-3. This is only one of twelve possible 3×3 Latin squares, and the actual placing of the plants was done at random.

OTHER VARIATIONS

Agricultural statisticians have developed a vast list of modified experimental designs, such as Partially Balanced Incomplete Block Designs and Lattice Designs. Most of these are highly specialized, however, and it is unlikely that they would be of utility to the ordinary laboratory biologist. The experiment on the growth of sugar beets in a plant growth room is very similar to an agricultural experiment, of course. Similar designs can be adapted to strictly laboratory experiments, such as the effects of several treatments on muscle contraction or the metabolism of cells, but the analogies to blocks, replicates, rows, and columns are often abstruse. Frequently single simply designed experiments go so rapidly that it is not worth the effort to use one of the abstract designs.

ANALYSIS OF VARIANCE

In the experiment described above, more than two populations are being described simultaneously. Differences between sample means can arise because the populations are different but also from chance alone. Variances are more difficult to estimate. Our experiment can be tested most effectively by means of analysis of variance. The following assumptions will be made: the samples are random, the variability is distributed according to the normal curve, and the variances of the different populations are equal. Again we shall use the hypothesis that the three treatments produce equal results, or that all the plants belong to the same population, and then we shall find the probability that the observed differences could arise from chance alone. Since the randomized block is used more frequently than the other designs, that one has been chosen for analysis.

After a period of growth, the plants of Fig. 17-2 were harvested, and the leaves were weighed. The weights are recorded in Table 17-2. There are three sample means, $\bar{x}_s$, and these are calculated at the bottom of the table. The working formula.

$$s^2 = \frac{\Sigma\, x^2 - (\Sigma\, x)^2/n}{n - 1}$$

is used for computing variance, and the various parts of this calculation are also included in the table. The value, $\Sigma\, x^2 - (\Sigma\, x)^2/n$, will be called the "sum of squares."

In this experiment there are three ways of estimating the variance of the population. The first is the variance s^2 computed over the entire experiment of twelve plants. The second consists of the variance within each of the treatments; for example, the four S plants have a variance. The third estimate is derived from the differences among the various treatments, $S, V,$ and L. Ultimately, if the variability among treatments is greater than the variability which can be attributed to chance, we must conclude that the treatments had some effect.

Referring to Table 17-2, an estimate of variance is obtained from the individuals in each treatment. The sums of squares are pooled, or added together, to yield a "total sum of squares." The variance of each treatment would be

$$s^2 = \frac{\Sigma\, x^2 - (\Sigma\, x)^2/n}{n - 1}$$

and the pooled value becomes

$$s^2 = \frac{\text{the sum of the three "sums of squares"}}{(n_S - 1) + (n_V - 1) + (n_L - 1)}$$

TABLE 17-2   *results of a randomized block experiment*

block	weight of leaves in grams			entire experiment
	S	V	L	
1	136	140	135	
2	143	151	128	
3	127	133	120	
4	137	142	131	
$\Sigma\, x$	543	566	514	1623
$\bar{x}$	135.75	141.50	128.50	135.25
$\Sigma\, x^2$	73843	80254	66170	220267
$(\Sigma\, x)^2/n$	73712	80089	66049	219511
$\Sigma\, x^2 - (\Sigma\, x)^2/n$	131	165	121	756
$n = 4,\ 3n = 12$				

The sum $(n_S - 1) + (n_V - 1) + (n_L - 1)$ is the number of degrees of freedom assigned to the individuals within treatments.

The means of the three treatments form another estimate of the population variance. The variance of this distribution of means is an estimate of the variance σ^2/n corresponding to the standard error $\sigma/\sqrt{n}$. The variance of the distribution of means is

$$\frac{(0.50)^2 + (6.25)^2 + (6.75)^2}{2} = 42.43$$

This is an estimate of $\sigma^2/4$, so the estimate of σ^2 is equal to 169.7. Since we are dealing with three values, there are two degrees of freedom. Multiplying by the number of degrees of freedom gives us a "sum of squares" which will be useful as a numerical check later.

TABLE 17-3   *analysis of variance of data in table 17-2*

source of variation	degrees of freedom	sum of squares	mean square
individuals	9	417	46.3
treatments	2	339	169.7
total	11	756	68.7

Now these figures are placed in an "analysis of variance" table (Table 17-3). Most of the numbers in this example are transferred directly from Table 17-2 or the foregoing discussion. The next step is to compare the variance estimates (called mean square in the table) by calculating a ratio called F.

$$F = \frac{\text{mean square of sample means}}{\text{mean square of individuals}} = \frac{169.7}{46.3} = 3.67$$

The table of distribution of F values shows us that a value of F this large occurs by chance more than 5 per cent of the time. In other words, the variations in the treatments are not large enough to be statistically significant, and it made no difference whether the plants were raised in soil, vermiculite, or liquid culture.

If the difference had been significant, further calculations, pinpointing the treatments which differed significantly would be possible.

FACTORIAL EXPERIMENTS It is often desirable to know the effects of several different factors on the responses of biological materials. Each of these factors could be tested separately by holding all the other factors constant. Temperature and the concentration of glucose both might influence the respiration of yeast cells, for example. Separate measurements of rate could be made at five different temperatures at any one level of glucose, and if these are replicated five times, there will be a total of twenty-five measurements. If we repeat this operation with each of five concentrations of glucose, we perform a total of 125 measurements. A factorial experiment can allow us to obtain the same information more easily and, at the same time, to evaluate any interaction between temperature and glucose concentration. At 5° C, we make five measurements, each using a different concentration of glucose. At 10°, 15°, 20°, and 25° C we do the same. We have performed a total of twenty-five measurements, and each temperature and each glucose concentration is replicated five times. Analysis of variance tells us the significance of the effect of temperature, the effect of glucose, and any change in the effect of temperature resulting from changes in the glucose concentration.

The factorial design is very logical. If a repetition of a measurement is identical in all details to the original measurement, the same errors are likely to be repeated. If several factors influence the results, one might as well learn something about these other effects upon repeating the measurement.

If the number of factors is large, the factorial design can become exceedingly complex. Because some of the factors may prove to be unimportant, the complex design may provide a great deal of useless information. A few preliminary measurements may simplify the design. Despite its complexity, the factorial design is very useful, and it is likely that more and more laboratory experiments will be performed in this manner.

I must repeat that a competent biometrician should usually be consulted during the experimental design phase of the research project.

SELECTED   Cochran, William G., and Gertrude M. Cox, *Experimental*
REFERENCES   *Designs,* 2nd ed. New York: John Wiley & Sons, Inc., 1957.
One of the most complete catalogues of experimental designs.

Steel, Robert G. D., and James H. Torrie, *Principles and
Procedures of Statistics.* New York: McGraw-Hill Book
Company, Inc., 1960. Better treatment of experimental
design than most general statistics textbooks.

You have just completed a research project. It is not really finished, of course, because the questions you have answered have only pointed out new questions, but you have solved the problem that you set out to investigate. It is only natural to feel a little proud because you have learned something about a little fragment of the universe that no one knew before. Science will be better off if you share your findings.

Experiments have been performed, measurements have yielded numbers, computations have been completed, and considerable thought has been given to interpretations. The next step is to prepare the material for publication. The preparation of the manuscript must follow the rules of the editor of the journal, so probably the first step is to decide in which journal your results should be reported. There are several thousand journals to choose from, but you know by now that only a selected few pertain to the specialized field in which you have been working. Among these, select on the basis of circulation, time required for publication, and the likelihood that the desired audience will be reached. Some scientific periodicals are sponsored by societies and accept papers only from members. The editor (or the editorial board) has prepared a set of rules, and every issue carries these rules or a note telling in which issue the rules were published. Only after taking these preliminary steps is it wise to proceed to prepare the report.

ORGANIZATION OF THE PAPER

The scientific paper more-or-less naturally falls into several main sections, although there are frequent modifications of these. Usually there is an introductory section, a description of experimental methods, a presentation of the results, discussion of interpretations, and a list of cited literature. Sometimes these sections are identified by headings; other times the text runs continuously from beginning to end. There may be an abstract at the beginning or end.

INTRODUCTION

The introductory statement tells the reader what to expect. It provides a specific statement of the problem and outlines other recent work on the same or similar problems. Several references to the literature probably will be required in order to place this paper in a proper relationship to what is already known.

EXPERIMENTAL METHODS Methods should be described in sufficient detail to allow others to repeat the work if they choose. What kind of organisms were used and how were they prepared for the experiments? If commercial instruments were used throughout, naming them is enough. If you developed any new techniques, however, these should be described in detail.

The description of the materials and methods must be a compromise. The journal cannot afford to publish long accounts of the method of holding a pipette, but any detail of technique which would not be obvious to a person experienced in this field must be described. If a previously used method was adopted in the current experiments, a reference to the previous paper can save words or paragraphs. A naked reference, however, may be unsatisfactory. For example, "Oxygen exchange was measured as described previously (with reference) . . ." tells the reader nothing, but "Oxygen was determined manometrically (with reference) . . ." may save him a trip to the library.

RESULTS Since most research is now quantitative, the results are expressed in numbers. The results can be presented in tables or graphs, but usually the journal cannot allow the same results in both forms. The text of this section carries a running description of the results to help the reader understand the tables or graphs.

DISCUSSION The results mean nothing unless they are related to the problem. The conclusions should answer the questions stated in the introduction. Perhaps the conclusion drawn in this paper can be applied to more general situations, and such generalizations can be pointed out.

LITERATURE CITED In several of the earlier sections you will have referred to the work of others. It is actually only polite to give credit to them and to tell your reader how he can find their papers. The list of references need not include every paper ever written on the subject, but it should include those papers needed by the reader to place your paper in the proper perspective.

THE JOB OF WRITING Some persons can sit down with only a headful of ideas and write a paper from beginning to end, but most of us need to follow a fairly detailed outline.

I suspect everyone develops his own system for writing a scientific paper. The purist insists that the writing start at the beginning, but this may not always be the easiest way. It often is easiest to write the section on materials and methods first because this section will be changed relatively little later. Probably the section on results can logically follow next. The introductory statement and the discussion of interpretations must cover the same subject matter, so it is best to write them together. Unless there is a careful correlation of these two sections, it is too easy to write conclusions to some problem other than the one investigated. The order in which the sections are written is a personal matter, and each writer learns by experience which order works best for him.

LITERARY STYLE   The main purpose of all writing is to communicate ideas. A paper is prepared, not to prove that the writer is a great scholar, but to convey facts and ideas to a reader. Clarity of presentation is just as important as accurate information and logical arguments. If I could give only one rule to a prospective writer, that rule would be, "Remember your reader!" Your subject has been foremost in your mind, and you may forget that what has become obvious to you is not obvious to everyone. The English language is expressive because of its built-in ambiguities, but in scientific exposition the intention of the author should be absolutely clear.

Scientific writing must be grammatically correct. Within the scope of correct English grammar there is plenty of room for an individual writing style. Simple, straightforward writing is the only kind suitable for science. The ideas of science are sufficiently complex without the distractions of overly dramatic or obscure expressions. Much bad writing has been included in the scientific literature, but progress is being made toward improvement. Unfortunately many beginning writers imitate what they read and so, commonly, learn bad habits.

One gains the impression from reading the literature that all verbs must be expressed in the passive voice to avoid saying, "I," but this convention is not necessary. The avoidance of the first person pronouns was once thought to add objectivity, but just how "It was discovered that . . . ." is less subject to personal bias than "I found . . . ." is a mystery. By whom "it was discovered" may even be a mystery to the reader. If the sentence can be written with active verbs, in the ordinary "subject, verb, object" order, by all means do so. No one can justify an awkward sentence by pointing out that the literature is full of such awkward sentences. Cast an occasional sentence backward when necessary or just for variety if you like.

Many scientists use what is sometimes called the "German construction," a compounding of nouns, adjectives, and verbs. As a simple example, examine "water regulating valve." Translated into German, there is no doubt about the meaning. In English, however, one is not sure if the water regulates the valve or vice versa. The confusion becomes worse if the object is a "solid brass solenoid controlled water influx and efflux regulating valve." Usually such constructions can be avoided by rearranging the sentence, without necessarily making the sentence longer.

Wordiness is probably one of the worst offenses in these days of crowded journals. The use of unnecessary repetition or of phrases that contribute nothing to the communication of ideas may even make the paper more difficult to read. One may say, "Concerning the cell membrane it must be kept in mind that it is of the utmost importance due to the fact that it serves the function of being responsible for the control of the movement of materials into the cell," but it would be better to say, "The cell membrane is important because it controls the movement of materials into the cell," or to state the probable intended idea more simply, "The cell membrane controls the movement of materials into the cell."

The jargon used in daily conversations in the laboratory has no place in the scientific paper. Biologists and biochemists deal with long and complex terms, and it is only natural to contract them into a slang. Your laboratory colleague knows perfectly well that the "Beckman" is the Beckman Model DU Spectrophotometer, but your reader might think of a Beckman *pH* meter. It is easy to invent words by adding -ate to indicate the result of a process, as in filtrate, or even "washate." Verbs can be concocted from adjectives by adding -ize (solubilize or counter-currentize). One of the most ridiculous jargon expressions I have read recently was a reference to the capacity of a technique as the "throughput."

Biochemists habitually use initials, but some of these have become so firmly established in the literature that their use is acceptable. Almost everyone knows ATP even if they do not remember that it stands for (A)denosine (T)ri(P)hosphate.

A few standard abbreviations of this sort are likely to be perpetuated, but certain ones should be avoided, because they are ambiguous. For example, TCA might be "trichloro-acetic acid" or "tricarboxylic acid." Deoxyribonuclease may be too long to pronounce, but there is no excuse for *writing* DNase. Most editors now recommend a minimum of such abbreviations and insist on their proper identification the first time they appear in the paper. Jargon probably cannot be avoided in laboratory speech, but it should not be allowed to creep into formal scientific writing.

A fault which is like jargon, in a way, is the use of technical terms where they are not needed. You might say, "It is immediately obvious to the observer that this specimen of *Canis familiaris* L. exhibits a predominantly melanistic pigmentation pattern," or you might say, "This dog is black." Think of yourself as the reader and take your pick.

SOME TECHNICAL DETAILS

You will need to refer to several earlier papers, and the actual form of the citations depends upon the journal. Some journals place all references in a list at the end of the paper, but others use footnotes. References are identified in the text material by Arabic numerals (superscripts or enclosed by parentheses) or by the name of the author and the year; for example, "Green (1968) showed . . . ." The references at the end of the paper might be listed in order of their appearance in the text, alphabetically by author, by year, or in some other order.

The citation itself also depends upon the journal. The following imaginary example illustrates the system I prefer:

Drake, P. D., and P. A. Mason. 1959. Effect of gamma rays on the yield of peaches. J. Agr. Biophys. 43:17–28.

Only the first word and proper nouns in the title are capitalized.

Unfortunately each journal uses its own system for citations and there is no uniformity. Common variations include inverted order for all author names, placing the year in some other position, omission of the title, and different systems of punctuation or different type styles. The abbreviations of the titles of periodicals should follow a system, preferably that used by *Chemical Abstracts*. Generally, examining a few recent issues of the selected journal helps one to learn the proper form for citations.

Tables and figures (graphs or other drawings) are the
common forms for presenting data.

The preparation of illustrations requires a little knowledge
about the printing processes. The original photograph or
drawing must be reproduced by means of an engraving. The
drawings furnished to the engraver usually are done on white
paper with black India ink. Because reduction occurs in
processing, the thickness of lines, the sizes of points on
graphs, and the proportions of the lettering must be chosen
to show clearly on the smaller picture. Some editors prefer
good glossy photographs of the original drawings. Graphs
may be drawn on coordinate paper (graph paper) if the
lines on the paper are blue, which does not photograph.

Most of the figures in this book were drawn in India ink
on $8\frac{1}{2}$ by 11 inch sheets of tracing paper. Lettering was done
with a Leroy lettering set, a device in which the pen follows
a prepared template. Several sizes of templates and pens are
available, and the right combination must be chosen for
the anticipated reduction in size of the printed drawing.

Graphs are often a problem for the beginner. We might
measure the effect of temperature on the rate of a reaction,
and the results would be especially amenable to graphic presentation. We control the temperature and measure the rates,
so temperature should be placed on the horizontal axis and
rate on the vertical axis. Some other sets of variables are not
quite so obviously "independent" or "dependent." Mathematical examination shows that one is a function of the
other, however, or $y = f(x)$. The length of a growing animal
varies with time, not vice versa. The absorption of light by
a solution depends on concentration, not the other way
around. Either you control and vary the factor and measure
its effect or you observe the effect of some uncontrollable
factor like time. In either case the effect goes on the vertical
axis.

The scale of distances on the graph corresponds to
numerical intervals. If 1 cm stands for 1 min, then 10 cm
stands for 10 min, even if measurements were made at 1 min,
2 min, 3 min, 5 min, 10 min, 20 min, or other odd intervals.
Figure 18-1 shows a properly prepared graph, as well as
some graphs containing common errors. Imagine how these
errors would affect the conclusions that might be drawn.

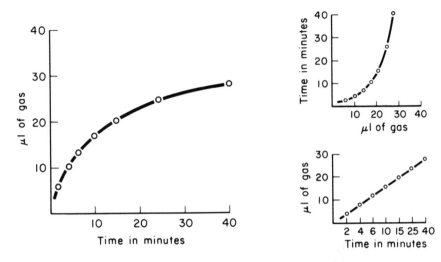

FIGURE 18-1   *Left, a properly prepared graph of the results of a typical experiment. The first few minutes were the most interesting, so readings were made at shorter intervals. Right, the same data: upper, axes interchanged; lower, time scale not linear.*

TABLES   A table is the most efficient way of presenting numbers. The table classifies the data, and its organization should be self-evident. Because the numbers in a table usually could be arranged in several different ways, it is desirable to try several forms and select the one that will demonstrate the results to the reader most effectively. Table 18-1 presents the same figures as Table 17-3, but in a different arrangement. If the two are compared, Table 17-3 is preferred because additions of vertical columns are easier than addition across a page.

TABLE 18.1   *data on analysis of variance from table 17-3*

source of variation	individuals	treatments	total
degrees of freedom	9	2	11
sum of squares	417	339	756
mean square	46.3	169.7	68.7

TYPING THE
MANUSCRIPT
The manuscript should be typed on good white paper, preferably 16-pound or 20-pound bond. If the surface is too rough, erasures are impossible; if too smooth or glazed, ink will smear. Use a typewriter with sharp, clean type faces and a ribbon in good condition.

Two copies are sent to the journal; therefore, you will need at least two carbon copies so that you may keep one yourself. Type on one side of the paper and double-space everything. Leave adequate margins (1 to $1\frac{1}{2}$ in.) for editorial comments and instructions to the printer.

Check the entire manuscript for typographical errors and errors in spelling, grammar, or content. Any corrections made during editing or proofreading somehow are quite likely to lead to new errors. If your manuscript is as nearly perfect as you can make it, your subsequent work of proofreading will be easier. Set the manuscript aside for a few days and then check it again. Let someone else, say a person not too familiar with your work, read it and criticize it. Select the readers carefully. If a reader says everything is "just fine," he must not have read it critically or he does not want to hurt your feelings. If another reader cannot find anything good about the manuscript and hands you a list of nasty comments, suspect him also. Eventually, after a few such trials, you will find two or more friends, relatives, or colleagues who can give you the kind of criticism you need. The purpose of the criticism, after all, is to improve the paper. The good critic will question something that is not clear to him; you had better rewrite that sentence or paragraph because you cannot explain it verbally to every reader.

Minor corrections can be corrected in ink or by typing on the manuscript. Major corrections may require retyping a page. Probably if the criticism was adequate you will want to retype the whole manuscript. After all, the scientific paper is usually only a few pages long, and the final typing is a small part of the entire job.

TITLE AND
ABSTRACT
These parts have been saved until last because they are the most difficult to write. Since a maximum amount of useful information must be included in a minimum of words, the words must be used very efficiently.

The title is the first item read by a reader searching the literature. Unless your title tells him what is in your paper, he may not read it, even though your findings might be important to him. Abstracting and indexing services frequently pick key words from titles. The title should tell the nature of the study, usually the approach to the problem, and often the kinds of organisms used. "The effects of gamma rays on the yield of peaches," mentioned earlier, provides enough information for a reasonable guess about the contents of the paper.

The abstract is a one paragraph summary, limited to about 200 words. Any reader is likely to read the abstract before he reads the whole paper, even if he has found the title appealing. The title is not repeated in the abstract, but the two complement each other. Describe the problem, the methods, the results, and the conclusions or interpretations. Point out any especially significant new findings. To do all this in 200 words requires a great deal of writing, editing, rewriting, pruning, and rearrangement. Words must be used more accurately and efficiently than in almost any other kind of writing.

PROBLEM   1. For practice in the kind of "tight" writing required for abstracts, try this exercise. Select from an issue of *Science* a "Report" with a title that appeals to you. Cover up the author's abstract at the beginning. Read the paper carefully and then write your own abstract. Check against the author's abstract to see how well you have picked out the most important ideas and how efficiently you have used your words.

SELECTED REFERENCES   Conference of Biological Editors, Committee on Form and Style. 1964. *Style manual for biological journals,* 2nd ed. Am. Inst. Biol. Sci. Washington. 92 pp. This manual has been prepared with the hope of standardizing and improving biological writing. It includes useful hints for improving style, together with information about the actual preparation of the paper. Every biologist should have a copy and *use* it.

Woodford, F. Peter, ed. *Scientific writing for graduate students.* New York: The Rockefeller University Press, 1968. A Council of Biology Editors manual that should be a great help even to experienced writers. Extensive samples of faulty text and improved text are presented side by side.

University of Chicago Press. 1969. *A Manual of Style,* 12th edition, revised. The Chicago manual has become one of the most widely-used reference manuals. In a number of instances it has become the authority governing writing and publishing in a variety of subjects.

Dictionary. An up-to-date, authoritative dictionary is the indispensable tool of the writer *and* the reader. Webster's New International Dictionary is the usual authority, but some of the smaller books may be more convenient on the desk.

BIBLIOGRAPHY

Suggested guides for laboratory work:

Bradshaw, L. Jack, *Introduction to Molecular Biological Techniques.* Englewood Cliffs, N. J.: Prentice-Hall, Inc., 1966. Each chapter consists of two parts, a theoretical introduction and a program of experimental work.

Dunn, Arnold S., and Joseph Arditti, *Experimental Physiology: Experiments in Cellular, General, and Plant Physiology.* New York: Holt, Rinehart and Winston, Inc., 1968.

For a background in biology:

Baldwin, Ernest W., *Dynamic Aspects of Biochemistry,* 5th ed. New York: Cambridge University Press, 1968. A thoroughly readable account of this important field.

Blackburn, Robert T., ed., *Interrelations: The Biological and Physical Sciences.* Chicago: Scott, Foresman and Company, 1966. A well-chosen collection of essays, classics, and classics-to-be.

Giese, Arthur C., *Cell Physiology,* 3rd ed. Philadelphia: W. B. Saunders Company, 1968. The physiology of cells, with no major distinction between animal and plant cells.

Hill, J. Ben, Henry W. Popp, and Alvin R. Grove, Jr., *Botany,* 4th ed. New York: McGraw-Hill Book Company, Inc., 1967. As a general reference on plants, this book is very useful.

McElroy, William D., and Carl P. Swanson, eds., *Prentice-Hall Foundations of Modern Biology Series,* 2nd ed. Englewood Cliffs, N. J.: Prentice-Hall, Inc., 1964. Each of these small books was written by a carefully selected, competent author. Taken together, they give a modern approach to the field.

Moore, John A., *Principles of Zoology.* New York: Oxford University Press, 1957. One of several good zoology texts.

Ramsay, J. A., *The Experimental Basis of Modern Biology.* New York: Cambridge University Press, 1965. Using selected experimental topics to illustrate biological principles.

Stanier, Roger Y., Michael Doudoroff, and Edward A. Adelberg, *The Microbial World,* 2nd ed. Englewood Cliffs, N. J.: Prentice-Hall, Inc., 1963. A complete textbook of microbiology, useful as a general reference because it includes algae, protozoa, and various fungi as well as bacteria. In fact, so many general principles of biochemistry, cellular physiology, and ecology are discussed, that this is one of the best references in general biology.

Advanced work relating to experimental research:

Kay, R. H., *Experimental Biology: Measurement and Analysis.* New York: Reinhold Publishing Corporation, 1964. Physical techniques used in physiology, biochemistry, and biophysics; strong section on electrical measurements.

Lenhoff, Edward, *Tools of Biology.* New York: The Macmillan Company, 1966. A paperback describing "tools" and advances made by using them.

Newman, David W., *Instrumental Methods of Experimental Biology.* New York: The Macmillan Company, 1964.

Richards, James A., Francis Weston Sears, M. Russell Wehr, and Mark W. Zemansky, *Modern University Physics.* Reading, Mass.: Addison-Wesley Publishing Company, Inc., 1960. A high-level complete introduction to modern physics.

Setlow, R. B. and E. C. Pollard, *Molecular Biophysics.* Reading, Mass.: Addison-Wesley Publishing Company, Inc., 1962. Generally excellent treatment of physical approaches to biological problems.

Strobel, Howard, *Chemical Instrumentation.* Reading, Mass.: Addison-Wesley Publishing Company, Inc., 1960. Complete discussions of instrumental analysis are included. A favorite of biologists as well as chemists.

Willard, H. H., L. L. Merritt, Jr., and John A. Dean, *Instrumental Methods of Analysis,* 4th ed. Princeton, N. J.: Van Nostrand Reinhold Company, 1964. The practical descriptions of most of the commercial instruments are quite detailed, and discussions of theoretical aspects are relatively easy to follow.

Wilson, E. Bright, Jr., *An Introduction to Scientific Research.* New York: McGraw-Hill Book Company, Inc., 1952. The single most instructive volume on experimental research, using examples from biology as well as from the physical sciences.

Sources of specific information:
FASEB Handbooks. Several handbooks of biological data published by the Federation of American Societies for Experimental Biology, Bethesda, Md., including the following: *Blood and Other Body Fluids,* 1961; *Growth,* 1962; *Biology Data Book,* 1964; and *Environmental Biology,* 1966.

Gray, Peter, ed., *Dictionary of the Biological Sciences.* New York: Reinhold Publishing Corporation, 1967.

Handbook of Chemistry and Physics. Cleveland: Chemical Rubber Publishing Co. New editions appear frequently; the most recent I have seen is the 50th, 1969.

McGraw-Hill Encyclopedia of Science and Technology, 2nd ed. New York: McGraw-Hill Book Company, Inc., 1966.

Williams, Roger J. and Edwin M. Lansford, Jr., eds., *The Encyclopedia of Biochemistry.* New York: Reinhold Publishing Corporation, 1967.